T0074786

Artificial Intelligence

Artificial Intelligence
Fundamentals and Applications

Edited by
Cherry Bhargava and
Pardeep Kumar Sharma

CRC Press
Taylor & Francis Group
Boca Raton London New York

CRC Press is an imprint of the
Taylor & Francis Group, an informa business

First edition published 2021
by CRC Press
6000 Broken Sound Parkway NW, Suite 300, Boca Raton, FL 33487-2742

and by CRC Press
2 Park Square, Milton Park, Abingdon, Oxon, OX14 4RN

CRC Press is an imprint of Taylor & Francis Group, LLC

ISBN: 978-0-367-55970-0 (hbk)
ISBN: 978-0-367-55973-1 (pbk)
ISBN: 978-1-003-09591-0 (ebk)

Typeset in Times
by codeMantra

Contents

Preface

Artificial intelligence (AI) is a science of making machines do things that would require intelligence if done by humans. AI is when machines can learn and make decisions similarly to humans. There are many types of AI, including machine learning, where instead of being programmed what to think, machines can observe, analyze, and learn from data and mistakes just like our human brains can. This technology is influencing consumer products and has led to significant breakthroughs in healthcare and physics as well as altered industries as diverse as manufacturing, finance, and retail. In part due to the tremendous amount of data we generate every day and the computing power available, AI has exploded in recent years.

This book consists of chapters that narrate the advances in AI along with its applications under common theme "Artificial Intelligence: Fundamentals and Applications." This book serves as a starting point for understanding how AI works with the help of real-life scenarios. By the end of this book, you will have understood the fundamentals of AI and worked through a number of real-life examples that will help you develop the intelligent vision. This book explores the potential consequences of AI and how it will shape the world in the coming years.

AI is the field of study that simulates the processes of human intelligence on computer systems. These processes include acquisition of information, using the information, and approximating conclusions. The research topics in AI include problem-solving, reasoning, planning, natural language programming, and machine learning. Automation, robotics, and sophisticated computer software programs characterize a career in AI. Basic foundations in maths, technology, logic, and engineering can go a long way in kick-starting a career in AI.

AI is a life-saving technology. From figuring out personalized drug protocols to better diagnostic tools and even robots to assist in surgeries, AI is altering our healthcare system from its processes to the care that these organizations provide. In the *future, artificial* limbs may even become stronger, faster, and more efficient. In addition to our healthcare systems, AI is likely to be very instrumental in solving the environmental concerns we are experiencing due to global warming.

This book takes its readers to the heart of the latest AI-thought process to explore the next phase of human existence. This book gives you a glimpse into AI and a hypothetical simulation of a living brain inside a computer.

Preface

Editors

Dr. Cherry Bhargava received the B.Tech. degree in EIE from Kurukshetra University; M.Tech. degree in VLSI design and CAD from Thapar University; and the Ph.D. degree in ECE, specialization in Artificial Intelligence, from State University, I. K. Gujral Punjab Technical University, Punjab. She is currently working as an Associate Professor, Department of Computer Science and Engineering, Symbiosis Institute of Technology, Pune, India. She has more than 16 years of teaching and research experience. She is GATE-qualified with All India Rank 428. She has authored about 50 technical research articles in SCI, Scopus-indexed quality journals, and national/international conferences. She has 18 books to her credit. She has registered five copyrights and filed 21 patents. Her four Australian innovative patents are granted. She is a recipient of various national and international awards for being outstanding faculty in engineering and excellent researcher. She is an Active Reviewer and an Editorial Member of numerous prominent SCI and Scopus-indexed journals. She is a guest editor with *Computer System Science Engineering Journal* and *World Journal of Science and Technology.*

Dr. Pardeep Kumar Sharma received the postgraduate degree in applied chemistry from GNDU, Amritsar, and the Ph.D. degree from Lovely Professional University. He is currently working as an Associate Professor with Lovely Professional University, India. He has more than 15 years of teaching experience in the field of applied chemistry, artificial intelligence, design of experiments, and reliability prediction. He has authored about 25 research articles in SCI, Scopus-indexed quality journals, and national/international conferences. He has six books to his credit, in the field of reliability and artificial intelligence. Four Australian patents are granted, 18 Indian patents are published, and two copyrights are registered. He is a recipient of various national and international awards. He is an Active Reviewer of various indexed journals. He is also a guest editor with *Computer System Science Engineering Journal.*

Contributors

Vibha Aggarwal
College of Engineering and Management
Punjabi University Neighbourhood Campus
Rampura Phul
Bathinda, Punjab, India

Kiran Ahuja
Department of Electronics and Communication Engineering
DAV IET
Jalandhar, Punjab, India

Shyam Bass
School of Pharmaceutical Sciences
Lovely Professional University
Phagwara, Punjab, India

Tusara Kanta Behera
School of Pharmaceutical Sciences
Lovely Professional University
Phagwara, Punjab, India

Anand Bewoor
Department of Mechanical Engineering
Cummins College of Engineering for Women
Pune, Maharashtra, India

Laxmi A. Bewoor
Department of Computer Engineering
Vishwakarma Institute of Information Technology
Pune, Maharashtra, India

Roshan Rajesh Bhakar
Department of Aerospace Engineering
Amity University Dubai
Dubai, UAE

Cherry Bhargava
Associate Professor
Department of Computer Science and Engineering
Symbiosis Institute of Technology, Pune
Cherry.bhargava@sitpune.edu.in

Shaveta Chugh
P.G. Department of Commerce
Khalsa College for Women
Civil Lines, Ludhiana, Punjab, India

Uddipta Das
School of Pharmaceutical Sciences
Lovely Professional University
Phagwara, Punjab, India

Amiya Kumar Dash
School of Engineering & Technology
BML Munjal University
Gurugram, Haryana, India

Sandeep Gupta
College of Engineering and Management
Punjabi University Neighbourhood Campus
Rampura Phul
Bathinda, Punjab, India

Tanu Kaistha
I.K.G Punjab Technical University
Jalandhar, Punjab, India

Prateek Kalia
Department of Corporate Economy
Faculty of Economics and Administration
Masaryk University
Lipova, Brno, Czech Republic

Mangesh Pradeep Kulkarni
School of Pharmaceutical Sciences
Lovely Professional University
Phagwara, Punjab, India

Ashwini Kumar
Department of Mechanical Engineering
National Institute of Technology
Jamshedpur, Jharkhand, India

Rajesh Kumar
School of Pharmaceutical Sciences
Lovely Professional University
Phagwara, Punjab, India

Ravinder Kumar
School of Mechanical Engineering
Lovely Professional University
Phagwara, Punjab, India

Tajunisa M.
School of Engineering,
Amity University
Dubai, UAE

A. Madana
School of Engineering
Amity University
Dubai, UAE

Astha Mishra
AKGEC-MCA
Ajay Kumar Garg Engineering College
Ghaziabad, Uttar Pradesh, India

Reshmi S. Nair
School of Engineering
Amity University
Dubai, UAE

L. R. Paul
School of Management Studies
Amity University
Dubai, UAE

Sesha Sai Kiran Poluri
Department of Engineering and Science,
University of Greenwich, Medway Campus
Central Avenue, Chatham
Kent, England

Priyanka
College of Engineering and Management
Punjabi University Neighbourhood Campus
Rampura Phul
Bathinda, Punjab, India

Ashutosh Rathore
School of Pharmaceutical Sciences
Lovely Professional University
Phagwara, Punjab, India

L. Sadath
School of Engineering
Amity University
Dubai, UAE

Sagar
School of Pharmaceutical Sciences
Lovely Professional University
Phagwara, Punjab, India

Anshu Sharma
Department of Computer Science & Engineering
CT University
Ludhiana, Punjab, India

Anurag Sharma
Department of Electronics and Communication Engineering
CT Institute of Technology and Research
Jalandhar, Punjab, India

Deepika Sharma
School of Pharmaceutical Sciences
Lovely Professional University
Phagwara, Punjab, India

Pardeep Kumar Sharma
School of Pharmaceutical Sciences
Lovely Professional University
Phagwara, Punjab, India

Rishav Sharma
School of Electronics and Electrical Engineering
Lovely Professional University
Phagwara, Punjab, India

Gurvinder Singh
School of Pharmaceutical Sciences,
Lovely Professional University
Phagwara, Punjab, India

Ranbir Singh
School of Engineering & Technology
BML Munjal University
Gurugram, Haryana, India

Udit Pratap Singh
University Institute of Computing
Chandigarh University
Mohali, India

Virinder Kumar Singla
College of Engineering and Management
Punjabi University Neighbourhood Campus, Rampura Phul
Bathinda, Punjab, India

Sarath Raj Nadarajan Assari Syamala
Department of Aerospace Engineering
Amity University Dubai
Dubai, UAE

Shalini Tripathi
School of Pharmaceutical Sciences
Lovely Professional University
Phagwara, Punjab, India

Yash Tyagi
School of Pharmaceutical Sciences
Lovely Professional University
Phagwara, Punjab

P. B. Vandana
School of Pharmaceutical Sciences
Lovely Professional University
Phagwara, Punjab, India

Jerrin Varghese
Department of Mechanical Engineering
The University of New South Wales
Sydney, Australia

Anurag Verma
School of Pharmaceutical Sciences
Lovely Professional University
Phagwara, Punjab, India

Nelvin Chummar Vincent
Department of Aerospace Engineering
Amity University Dubai
Dubai, UAE

Sheetu Wadhwa
School of Pharmaceutical Sciences
Lovely Professional University
Phagwara, Punjab, India

Pankaj Wadhwa
School of Pharmaceutical Sciences
Lovely Professional University
Phagwara, Punjab, India

1 Artificial Intelligence and Nanotechnology

A Super Convergence

Virinder Kumar Singla, Vibha Aggarwal, Priyanka, and Sandeep Gupta
College of Engineering and Management

CONTENTS

1.1 INTRODUCTION

Current technological and scientific progress is increasingly dependent on three technologies, namely, biotechnology, information technology, and nanotechnology. The idea of integrating bioscience, artificial intelligence (AI), and nanotechnology will foster yet another revolution in the field of science and technology, which has lingered for more than a decade. Nevertheless, the planned integration of multidisciplinary research is still developing.

Nanotechnology incorporates an understanding of engineering and physical sciences; it's one of the most important emerging technology sectors, and it is used in diverse areas such as medical, engineering, and agriculture.

AI is an approach to inculcate human-like thinking into an electronic gadget of any scale. This is an analysis of how human brain—as it attempts to solve problems—thinks, learns, decides, and works. It is heavily inspired by biological anatomy for the development of prevalent and most effective models, viz., artificial neural networks (ANNs) and other such algorithms. An important AI goal is to improve machine functions related to human intelligence, such as reasoning, thinking, and problem-solving. AI has been deployed in a steadily expanding set of fields: not only within itself, where the fields of machine learning, deep learning, and ANNs are now effective methods in their own right, but also in the number of areas and industries in which it now prevails. The adoption of AI together with the Internet of Things (IoT) and other developing sectors has already revolutionized many manufacturing and monitoring processes across the various industries, and is still growing.

Nanotechnology mostly comprises complex systems that are not always consistent with the different aspects of AI. However, it has been believed that AI will use nanotechnology as a tool to converge to oneness. Though such a vision still seems futuristic, a similar harmonization has started to unveil in the modern technology. A combination of these two fields can result in great breakthroughs from the fast-paced AI-assisted nanotechnology research, to creating the state-of-the-art materials, to expand the application area of AI using nanotechnology-based computing devices. Besides merging the two technologies, a combined research can also give a thrust to the study in each discipline, possibly leading to all sorts of new methods to gain insights and communication technologies. The "convergence" is subjected to being part of wider political and social discourses on biotechnology, cognitive study, nanotechnology, robotics, AI, information and communication technology (ICT), and the sciences concerned with such subjects (Figure 1.1).

Meanwhile, numerous initiatives have employed AI technologies in the field of nanoscience research, viz., to analyze the investigational methods and to aid in the design process of new nanodevices and nanomaterials. There are several reasons why nanoresearch uses the AI paradigms. Nanotechnology grieves from the instinctive boundaries of the scale of work; here, governing physical laws are altogether dissimilar from those otherwise applicable normally. Hence, the correct elucidation of the outcomes attained from any such system is one of the glitches that nanotechnology needs to address (Ly et al. 2011). To make things worse,

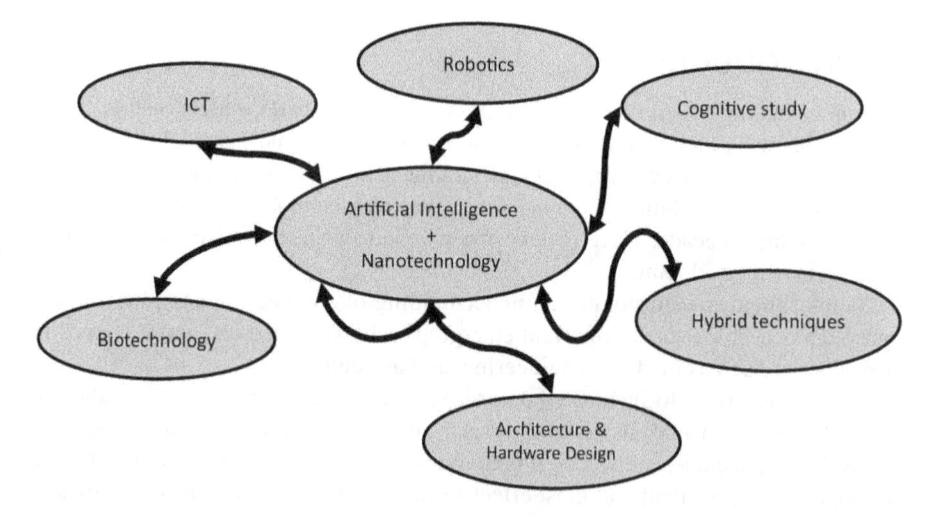

FIGURE 1.1 Artificial intelligence, nanotechnology, and other techniques converging together for the better.

several components in many systems influence the signal heavily. The development of theoretical approximations is challenging in these situations, and simulation techniques have been utilized to obtain precise elucidations of the investigational results. Here, various AI machine learning paradigms can become a handy tool in both generating research outcomes and developing nanoapplications in the future. Such techniques are very effective in dealing with several interrelated parameters in parallel and can well state and simplify complex/unknown data or functions (Mitchell 1997; Bishop 2006). Machine learning approaches such as ANNs, a set of weighed connected nodes, and the link weights are used to study these kinds of functions, using the monitored or unmonitored algorithm, which will be highly useful. Various optimization and search problems can be resolved by other AI techniques. There are many machine learning techniques that either involve single or a combination of methods, comprising decision trees, support vector machines, Bayesian networks, etc., that can be deployed in nanotechnology research for the multifaceted classification, prediction, correlation, data mining, clustering, and other control problems.

Also, a few studies have been carried out on how AI techniques could harness the computational power boost offered by future nanomaterials, developed by nanoscience, to be used for fabricating nanodevices, and nanocomputing will offer the powerful dedicated architectures for applying machine learning techniques.

The next section ponders upon this bidirectional relationship between AI and nanotechnology by means of various exemplary uses and applications.

1.2 UTILITY OF ARTIFICIAL INTELLIGENCE

1.2.1 AI in Scanning Probe Microscopy

Scanning probe microscopy (SPM) is the commonly used imaging technique in the nanoworld. Numerous strategies that obtain images by the interaction between a pattern and a probe fall beneath this concept. Characterization of the pattern topography is accomplished by using the tunneling current between the pattern and the probe through their interaction. Several techniques have been developed by varying interactions among the tip and the sample, after the invention of nanoscope. SPM is likewise an effective tool for an atomic-scale manipulation.

Challenges regarding the interpretation of the microscopic signals still remain, even though many efforts were made to enhance the decision and the capacity to manipulate atoms. The probe–sample interactions are not easy to apprehend and depend on many parameters. AI strategies may be an extraordinary rescue to solve such issues.

Further progress in the multimodal SPM imaging for obtaining extra complementary information (approximately the pattern), in recent times, produced a huge amount of information, therefore making it even more tough to interpret specific properties of the sample. To cope with this issue, a method has been developed called functional identification imaging (FR-SPM), which seeks a direct identity of local behaviors measured from spectroscopic responses and the usage of neural networks educated on examples provided by means of an expert.

Cellular genetic algorithm (cGA), a Gas subclass, is totally based on the evolutionary optimization algorithm, which is used to automate the imaging procedure in SPM with software capable of improving the precise state of the probe and the associated control parameters. Superior atomic resolution pictures are hence obtained with no human intervention except preparing samples and tips (Huy et al. 2009; Woolley et al. 2011).

ANNs are extensively used for the categorization of various behavioral, structural, and physical properties of nanomaterials on the nanoscale, which are used in plenty of applications, viz., CNT (carbon nanotube), quantum-dot semiconductor optics and devices, chemical technology, and production industry.

1.2.2 Nanosystem Design

Recently, ANNs have been used inside the transparent conductive oxide deposition process to determine the nonlinear relationship between input variables and output responses. This form of thin film has currently been used as an electrode in optoelectronic devices such as solar cells, organic LED, and flat-panel displays (Bhosle et al. 2006).

Evolutionary optimization was also used to develop better structures for nanoantennas, which outperform the best available radio-wave type of reference antennas. Using GA, the fittest antenna geometry suggests that it merges the characteristics of the fundamental magnetic resonance of the split-ring with the electrical one among the linear dipole antennas (Feichtner et al. 2012). This approach will refine nanoantenna architectures for special purposes and adequately deliver new layout techniques by carefully studying the operating principles of the resulting geometries.

GAs have also observed their uses inside the nano-optics field. In diverse nano-optics applications, such as optical manipulators, solar cells, plasmon-enriched photodetectors, modulators, or nonlinear optical devices, a vigilant design of nanoparticle mild concentrators will have a big impact.

1.2.3 Nanoscale Simulation

One of the major issues which scientists have to face when working at the nanoscale is related to the tool simulation being studied as actual optical pictures at the nanoscale cannot be achieved. Images must be interpreted at this scale, and numerical simulations are once in a while the best technique to get an accurate scheme of what is present in the image. Nonetheless, they are still tough to apply in many conditions, and lots of parameters need to be taken into account on the way to get a reasonable system depiction. Here, AI can be useful in enhancing the simulations' performance and making them simpler to collect and interpret.

The use of ANNs in numerical simulations has been proven to be beneficial in various approaches when operating at the nanoscale. First, the software program can be manually modulated to control the stability between numerical exactness and physical implication. Another use of ANNs in simulation software is to lessen the complexity of configuration related to them (Castellano-Hernández et al. 2012).

1.2.4 Nanocomputing

There is a vast diversity of applications that emerge from the mixture of AI and current and upcoming nanocomputing methods (Service 2001; Bourianoff 2003). AI paradigms have been used for the various levels of modeling, designing, and building prototypes of nanocomputing gadgets since the beginning of nano-computers. Machine learning tactics implemented with the aid of nano-hardware to a certain extent to semiconductor-based hardware can also provide a foundation for a new technology of less costly and transportable era that can comprise high overall performance computing, including programs, sensory facts processing, and control tasks (Uusitalo et al. 2011; Arlat et al. 2012).

The best expectations from the nanotechnology-enabled quantum computing and storage can considerably boost our capacity to clear up very complicated NP-whole optimization dilemma. Such sorts of issues arise in many unique contexts, but mainly those in Big Data that requires "computational intelligence" (Ladd et al. 2010; Maurer et al. 2012).

Natural computing normally takes place in distinctive techniques in this context. Techniques inclusive of DNA computing or quantum computing are properly studied at the present apart from other natural computing procedures being followed (Darehmiraki 2010; Razzazi and Roayaei 2011; Ortlepp et al. 2012; Zha et al. 2013).

In DNA computing, a lot of variables are in use. This is a scenario in which DNA computing AI strategies are useful for purchasing an ultimate result from a minor preliminary data set, preventing the usage of all candidate solutions. Evolutionary and GAs are another options that may be considered.

Eventually, the design of nanocomputing systems—few are bioinspired—includes a wide variety of emerging nanotechnologies. During the accumulation of recent physical working bases, reconfigurable architecture storage, and computational schemes, these technologies will be able to use new data version to apply machine learning paradigms in order to solve complex problems in a wide variety of applications.

1.3 FOOD SCIENCE

Food science is growing quickly in association with nanotechnology. The food market demands technology that is vital to hold marketplace leadership within the food processing industry in order to produce reliable, suitable, and tasteful fresh food products, and nanotechnology is the answer. Nanoparticles ("nano inside," "nano outside") are used as preservatives and wrapping, respectively.

Nanoscale food additives may be used to have an effect on product taste, nutrient composition, shelf life, and texture; may be used even to pick out pathogens; and may act as signs of meals' quality. Nanotechnology presents an enormous array of possibilities for the development of recent products and meals' system applications. AI strategies are pretty supportive for research and development possibilities for food additives and packaging.

1.4 NANOBOTS IN MEDICINE

It is established that a lot of nanosystems are functional for their interaction with alive neurons. For example, a few CNT properties allow us to lay out nanotube detectors that might help implement the pulse-train neural network behavior because the detectors are threshold gadgets similar to spiking neurons (Lee et al. 2003).

ANN techniques are lucratively used to examine the detection of unstable organic compounds by using CNT-covered acoustic and optical sensors. It is concluded that a first-rate categorization exchange is achieved through the aggregate of various modules of acoustic and optical sensors, which is the state of affairs where ANNs display their full potential (Penza et al. 2005).

Alternatively, ANNs have been considered as a well-known device for nanoparticle training analysis and modeling in the context of pharmacology and nanomedicine, with a high ability effect in chronic decease (Zarogoulidis et al. 2012).

Investigators at the University of California San Diego (UCSD) have designed nanobots capable of cleansing the blood of toxins generated by means of bacteria. These nanobots are about 25 times smaller than the width of a human hair and may travel 35 µm in step with second by way of "swimming" through the blood while powered with the aid of ultrasound.

Nanobots evolved with the aid of MIT researchers in 2018 are so tiny and light that they might float within the air. This nanotechnology can be made viable by means of linking 2D electronic additives to minute particles measuring between one billionth and one millionth of a meter. The final end result is a robot, which is no larger than an ovum or a grain of sand.

The merging of photodiode semiconductors, which have the capability to discover radiation from the optical area and transform it into an electrical signal, permits for a continuous supply of strength to the environmental sensors embedded in these robots. The small electrical fee generated is enough to permit this technology to perform without a battery.

As for the worth of these nanobots, the investigators plan to launch them on missions in remote locations to reveal surroundings together with pipelines and the human digestive system. This microscopic emissary can be released into the opening, allowed to flow through the direction of the pipe, and then recovered at the exit. Once harvested, the statistics collected via its sensors, consisting of the spatiotemporal attention of positive chemical substances like hormones and enzymes, may be downloaded and then considered.

1.5 SUMMARY

AI could efficiently offer solutions to many problems arising in the study of nanotechnology. It has been explored the use of ANNs and GAs in many exclusive contexts ranging from data interpretation in microscopy scanning probe to the characterization and category of nanoscale cloth properties. It has also reviewed various efforts to construct nano-machines and use them to realize the cutting-edge paradigms for synthetic intelligence. These pioneering efforts call for a real convergence of nanotechnology and AI in high-performance computer systems enabled by means of primarily biomaterial- based nanocomputing devices. Finally, it has been proven the extensive capacity impact on the use of AI strategies. However, nanotechnology has been used in biomedical study, remedial applications, and food science.

Nanotechnology deals with the bottom-up design, while AI research normally offers a top-down approach to trouble shooting The convergence of those fields will provide techniques to numerous complicated issues, which needs more than one layers of explanation and relations. Nanotechnology and AI can help on this effort, as mentioned above, which revitalize.

REFERENCES

Arlat, Jean, Zbigniew Kalbarczyk, and Takashi Nanya. "Nanocomputing: Small devices, large dependability challenges." *IEEE Security & Privacy* 10, no. 1 (2012): 69–72.

Bhosle, V., A. Tiwari, and J. Narayan. "Metallic conductivity and metal-semiconductor transition in Ga-doped ZnO." *Applied Physics Letters* 88, no. 3 (2006): 032106.

Bishop, Christopher M. *Pattern Recognition and Machine Learning*. Berlin: Springer (2006).

Bourianoff, George. "The future of nanocomputing." *Computer* 36, no. 8 (2003): 44–53.

Castellano-Hernández, Elena, Francisco B. Rodríguez, Eduardo Serrano, Pablo Varona, and Gomez Monivas Sacha. "The use of artificial neural networks in electrostatic force microscopy." *Nanoscale Research Letters* 7, no. 1 (2012): 1–6.

Darehmiraki, Majid. "A semi-general method to solve the combinatorial optimization problems based on nanocomputing." *International Journal of Nanoscience* 9, no. 5 (2010): 391–398.

Feichtner, Thorsten, Oleg Selig, Markus Kiunke, and Bert Hecht. "Evolutionary optimization of optical antennas." *Physical Review Letters* 109, no. 12 (2012): 127701.

Huy, Nguyen Quang, Ong Yew Soon, Lim Meng Hiot, and Natalio Krasnogor. "Adaptive cellular memetic algorithms." *Evolutionary Computation* 17, no. 2 (2009): 231–256.

Ladd, Thaddeus D., Fedor Jelezko, Raymond Laflamme, Yasunobu Nakamura, Christopher Monroe, and Jeremy Lloyd O'Brien. "Quantum computers." *Nature* 464, no. 7285 (2010): 45–53.

Lee, Ian Y., Xiaolei Liu, Bart Kosko, and Chongwu Zhou. "Nanosignal processing: Stochastic resonance in carbon nanotubes that detect subthreshold signals." *Nano Letters* 3, no. 12 (2003): 1683–1686.

Ly, Dung Q., Leonid Paramonov, Calvin Davidson, Jeremy Ramsden, Helen Wright, Nick Holliman, Jerry Hagon, Malcolm Heggie, and Charalampos Makatsoris. "The matter compiler-towards atomically precise engineering and manufacture." *Nanotechnology Perceptions* 7, no. 3 (2011): 199–217.

Maurer, P.C., G. Kucsko, C. Latta, L. Jiang, N.Y. Yao, S.D. Bennett, F. Pastawski, D. Hunger, N. Chisholm, M. Markham, and D.J. Twitchen. "Room-temperature quantum bit memory exceeding one second." *Science* 336, no. 6086 (2012): 1283–1286.

Mitchell, Tom M. *Machine Learning*. Maidenhead: McGraw Hill (1997).

Ortlepp, Thomas, Stephen R. Whiteley, Lizhen Zheng, Xiaofan Meng, and Theodore Van Duzer. "High-speed hybrid superconductor-to-semiconductor interface circuit with ultra-low power consumption." *IEEE Transactions on Applied Superconductivity* 23, no. 3 (2012): 1400104.

Penza, M., G. Cassano, P. Aversa, A. Cusano, A. Cutolo, M. Giordano, and L. Nicolais. "Carbon nanotube acoustic and optical sensors for volatile organic compound detection." *Nanotechnology* 16, no. 11 (2005): 2536.

Razzazi, Mohammadreza, and Mehdy Roayaei. "Using sticker model of DNA computing to solve domatic partition, kernel and induced path problems." *Information Sciences* 181, no. 17 (2011): 3581–3600.

Service, Robert F. "Nanocomputing. Assembling nanocircuits from the bottom up." *Science* 293, no. 5531 (2001): 782.

Uusitalo, Mikko A., Jaakko Peltonen, and Tapani Ryhänen. "Machine learning: How it can help nanocomputing." *Journal of Computational and Theoretical Nanoscience* 8, no. 8 (2011): 1347–1363.

Woolley, Richard A. J., Julian Stirling, Adrian Radocea, Natalio Krasnogor, and Philip Moriarty. "Automated probe microscopy via evolutionary optimization at the atomic scale." *Applied Physics Letters* 98, no. 25 (2011): 253104.

Zarogoulidis, Paul, Ekaterini Chatzaki, Konstantinos Porpodis, Kalliopi Domvri, Wolfgang Hohenforst-Schmidt, Eugene P. Goldberg, Nikos Karamanos, and Konstantinos Zarogoulidis. "Inhaled chemotherapy in lung cancer: Future concept of nanomedicine." *International Journal of Nanomedicine* 7 (2012): 1551.

Zha, Xinwei, Chenzhi Yuan, and Yanpeng Zhang. "Generalized criterion for a maximally multi-qubit entangled state." *Laser Physics Letters* 10, no. 4 (2013): 045201.

2 Artificial Intelligence in E-Commerce

A Business Process Analysis

Prateek Kalia
Masaryk University

CONTENTS

2.1 INTRODUCTION

Researchers argue that the fourth industrial revolution will be powered by information and communication technology, machine learning, digitization, robotics, and artificial intelligence (AI). Machines will be utilized to make decisions, thus creating a profound impact on business marketing practices and society (Syam and Sharma 2018; Dwivedi et al. 2019). In the next 20 years, the AI revolution will have an even greater impact than the industrial and digital revolutions combined (Makridakis 2017). Studies have confirmed that the emergence of intelligent products and services is not just hype as these possess the capability to transform the world (Sonia et al. 2020). Researchers are convinced with the genesis of AI, which has already crossed two "hype cycles," i.e., the first hype cycle in 1950–1983, resurgence in 1983–2010, and the second hype cycle in 2011–2017. Now AI is going to be the future of brains, minds, and machines (2018–2035) (Aggarwal 2018; Simon 2019). In a recent survey study, Frey and Osborne (2017) found that 47% of US jobs could be automated by 2033, as AI will have a significant impact on sales, marketing, and customer service.

Experts have predicted a substantial effect of AI in three industries, i.e., retail, education, and health care (Ostrom et al. 2018). Retail industry with a high proportion of human work and concurrent low-profit margins is a natural fit for AI applications, especially e-commerce (Weber and Schütte 2019). Commonly, e-commerce companies are using AI for personalizing websites and product recommendations (Netflix). Tech and retail giants such as Amazon.com are heavily investing on the research and development for advancing AI applications like recommendation engine (Alexa), voice-powered assistants (Echo), the Prime Air drone initiative, etc. In fact, Amazon is offering AI and machine learning capabilities to other companies through its cloud platform (AWS) (Weber and Schütte 2019).

To understand the role and importance of AI in e-commerce, this chapter will discuss AI and its applications in various business processes, involved in an e-commerce business. But before starting the discussion, we should first understand the concept of AI, which is detailed in the proceeding section.

2.2 ARTIFICIAL INTELLIGENCE

Before discussing AI, we should first understand "intelligence," which is the ability of a person to learn, understand, or deal with new situations; think abstractly; and use knowledge to manipulate one's environment (Merriam-Webster.com 2020). In general terms, intelligence can be defined as the ability to acquire and apply memory, knowledge, experience, understanding, reasoning, imagination, judgment, opinions, facts, skills, calculations, information, and language in order to calculate, classify, generalize, and perceive relationships; solve problems; plan and think abstractly; comprehend complex ideas; learn quickly; overcome obstacles; and adapt efficiently to new situations, either by changing oneself or by changing the environment (Legg and Hutter 2006; Paschen et al. 2019).

2.2.1 AI Mimicking Human Intelligence

The concept of AI originated on the considerations involving the extent to which the machine can partially or completely replace humans in the performance of tasks (Weber and Schütte 2019). Therefore, the marketing literature defines AI in terms of human intelligence. For example, researchers define AI as machine that exhibits the aspects of human intelligence (Huang and Rust 2018), mimics intelligent human behavior (Syam and Sharma 2018), or mimics as nonbiological intelligence (Tegmark 2017). Similarly, McCarthy (2007) defines AI as "the science and engineering of making intelligent machines, especially intelligent computer programs. AI is related to the similar task of using computers to understand human intelligence, but it does not have to confine itself to the methods that are biologically observable." These definitions make AI contingent on human intelligence (Bock et al. 2020).

2.2.2 AI Exceeding Human Intelligence

At times, humans indulge in behaviors that may not lead to the best final outcome (Kahneman and Tversky 1979), because of bounded rationality arising due to a limited information, cognitive abilities, and a finite time to make decisions (Dawid 1999). There are scientists who believe that machines can exhibit a human-like intelligence in two ways: acting intelligently (performing processes like memorizing, learning, reasoning, perception, and problem-solving towards a goal-directed behavior) and rationality (achieving "right thing" under uncertainty) (Paschen et al. 2019).

AI, with the help of Big Data and deep learning, can identify inclinations, intentions, and patterns that are beyond the intellectual capacity of a human brain. Human brain can interpret and conclude from the limited data; however, machines can interpret billions of data points. AI has advanced through "four intelligence processes" (i.e., from analytical to emotional) (Huang and Rust 2018) to acquire advanced capabilities like reasoning, planning, conceptual learning, creativity, common sense, cross-domain thinking, and even self-awareness (Bock et al. 2020). In this scenario, the definition of AI proposed by Kaplan and Haenlein (2019) sounds more appropriate, i.e., AI "as a system's ability to interpret external data correctly, to learn from such data, and to use those learnings to achieve the specific goals and tasks through a flexible adaptation."

2.3 E-COMMERCE BUSINESS PROCESSES AND ARTIFICIAL INTELLIGENCE

There are various business processes like marketing, buying, selling, and servicing of products and services performed by the companies involved in e-commerce (Figure 2.1). These businesses completely depend on e-commerce applications and internet-based technologies to carry out marketing, discovery, transaction processing, and product and customer services. E-commerce websites carry out interactive marketing, ordering, payment, and customer support processes on the World Wide Web. E-commerce also includes the processes related to e-business, where suppliers and customers access inventory databases through extranet (transaction processing), or sales and customer service representatives access the customer relationship management (CRM) systems via an internet (service and support), or customers collaborate in the product development through email and social media (marketing/discovery) (O'Brien and Marakas 2011).

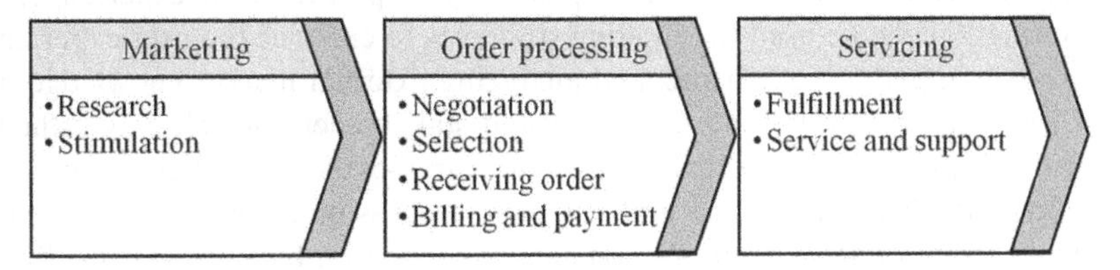

Source: Adapted from O'Brien and Marakas (2011)

FIGURE 2.1 E-commerce business processes supporting electronic selling of goods and services.

Researchers believe that AI can improve the business performance because the AI solutions are faster, cheaper, and less prone to human mistakes (Huang and Rust 2018; Canhoto and Clear 2020). Therefore, in the next sections, we will try to understand how AI is contributing to various e-commerce business processes.

2.3.1 Marketing

2.3.1.1 Market Research

Till the third industrial revolution, businesses used information technology for data processing and communication only. However, the fourth industrial revolution will allow computers to make appropriate and reliable decisions (Syam and Sharma 2018). Researchers believe that large businesses failing to deploy the latest technologies such as AI will be swept away in the face of competition (Stone et al. 2020). The primary focus of the market research is to identify the accurate segments of customer groups. Selected segments are targeted with suitable products, offered at appropriate prices, and supported with reasonable promotional and communication strategies to deliver products to customers through an appropriate distribution strategy (Syam and Sharma 2018). Earlier, segmentation was generally based on "traditional" techniques (like cluster analysis (for clustering), chi-squared automatic interaction detection (CHAID) (for classification)) or more recent segmentation techniques (like hidden Markov models, support vector machines, artificial neural networks (ANN), classification and regression trees (CART), and genetic algorithms). Machine learning tools of the new century have increased the efficiency of these segmenting algorithms. For example, the AI-empowered ANN models can find the solutions for marketing problems faced by B2B e-commerce companies (Wilson and Bettis-outland 2020). Marketers have enormous statistics in their hands to process a massive unstructured data (Big Data) for segmentation with the help of unsupervised neural networks (Hruschka and Natter 1999). Now profitability or customer lifetime value (CLV) segments can be identified based on the machine learning-powered decision trees (Florez-lopez and Ramon-jeronimo 2009). AI can process a huge amount of written and nonwritten user-generated contents available on social media platforms to reveal the user needs, preferences, attitudes, and behaviors. For example, IBM Watson AI system can identify the psychographic characteristics expressed in a piece of text to give valuable insights to marketers for the new product development or innovation (IBM 2020). AI can identify themes and patterns in users' posts to interpret the user experience, and the information can be used for creating strategies to enhance the user experience. It also helps in gathering, sorting, and analyzing external market knowledge, i.e., intelligence about external market forces and stakeholders, which may influence the customer preferences and behaviors. For example, the AI systems empowered by the machine learning and natural language processing algorithms can identify fake news from a huge amount of contents published on blogs, social media etc. (Berthon and Pitt 2018). Similarly, competitive intelligence can be developed by identifying themes or keywords from unstructured data (news, social media, website content etc.) (Paschen et al. 2019).

2.3.1.2 Market Stimulation

Market stimulation is concurrent to marketing, which is "the activity, set of institutions, and processes for creating, communicating, delivering, and exchanging offerings that have value for customers, clients, partners, and society at large" (American Marketing Association 2017). Typically, it encompasses four separable but interlinked components, i.e., product, price, place, and promotion (McCarthy 1960). However, the concept of marketing has arrived at an evolutionary point where an adaptation to technology is imperative and an impact of AI under each marketing mix component is obvious, i.e., for product (hyper-personalization, new product development, automatic recommendations etc.), price (price management, personalized dynamic pricing), place (convenience, speed, simple sales process, 24/7 chatbot support etc.), and promotion (personalized communication, unique user experience, creating wow factor, minimizing disappointment etc.) (Jarek et al. 2019; Dumitriu and Popescu 2020). Out of five AI areas (i.e., image recognition, text recognition, decision-making, voice recognition, and autonomous robots and vehicles), the first three are used quite extensively in marketing. Marketers are cautiously implementing AI applications because of cost and uncertainty attached with them. However, large tech companies like Google, Amazon, Microsoft, and Apple are investing on AI areas like voice recognition and autonomous robots and vehicle solutions (Jarek et al. 2019). As a result, there is a significant effect of AI on contemporary marketing practices; for example, routine, time-consuming, and repeatable jobs have been automated (data collection, analysis, image search, processing); strategic and creative activities to build a competitive advantage are emphasized; and business enterprises are designing innovative ways to deliver the customer value. It has further created a marketing ecosystem where entities offering AI solutions are in demand (Jarek et al. 2019).

2.3.2 Transaction Processing

2.3.2.1 Terms Negotiation

Negotiation can be defined as an art (more than science) to get what you want from bargaining or a person-to-person interaction. Most of the interactions in case of e-commerce transaction are electronic, i.e., through email, social media, text chat, or phone. In this setting, AI can be applied as a functional science to give an advantage in the negotiation process (Mckendrick 2019). Negotiation is like a persuasion dialogue where a series of arguments are proposed by a proponent and an opponent iteratively, and both issues counter arguments to defeat each other's argument (Huang and Lin 2005). Based on the components of negotiations (i.e., negotiation set, a protocol, a collection of strategies, and a rule of deal), scientists are training their chatbots and virtual assistants to plan several steps ahead and assess how saying different things could change the outcome of the negotiation (Reynolds 2017). These AI negotiation agents can operate 24/7 on behalf of e-retailer to locate customers and automatically negotiate to the best term as per the parameters set by the administrator or even market conditions (Krasadakis 2017). It is quite difficult for online stores to engage in a communication through graphical user interfaces (websites) to acquire, serve, and retain customers. Similarly, there is no chance for the customers to negotiate for a better deal

(Huang and Lin 2005). But deploying natural language interface for human–computer interaction can effectively solve this problem (Jusoh 2018).

E-commerce is highly dynamic market, and prices change rapidly. The E-commerce companies are using AI for dynamic pricing of their product and services, i.e., adjusting prices as per the market conditions (demand–supply) on a real-time basis (Kephart et al. 2000).

2.3.2.2 Order Selection and Priority

Recently, Alibaba launched Fashion AI technology to boost sales. Customers can upload pictures of a product they would like to buy to its Taobao e-commerce site, and the website automatically searches that item for sale similar to the photo (Simon 2019). Similarly, AI can collect the real-time data by tracking an online activity of the customer on their or competitor's website to decide and offer a price discount or search through company's database to check if those shoppers have rejected or accepted the previous product recommendations (Canhoto and Clear 2020). Another important application of AI is replenishment optimization. AI can reduce the inventory costs by determining the right time and quantity, for placing an order to the central warehouse and the suppliers (Stone et al. 2020). This can resolve several issues such as a reduction in number or amount of the unsold goods, optimization of the shelf space in warehouse, and an increase in cash flow. AI algorithms can optimize the individual order and delivery (personalization) (Zanker et al. 2019) and simplify the complex tasks like the same-day delivery (Kawa et al. 2018).

2.3.2.3 Order Receipt

With the help of predictive systems, AI can evaluate the prospects (customers) on their propensity to buy (high-quality leads) (Järvinen and Taiminen 2016), answer common questions and overcome objections of customers by using emotional AI (Paschen et al. 2019), and automate and speed up the checkout process (Campbell et al. 2020). Front-runner e-commerce companies like Amazon have introduced a language-assisted ordering (Amazon Echo) (Holmqvist et al. 2017). Complex AI models are used for sales forecasting (Dwivedi et al. 2019); store assortments (Shankar 2018); and personalizing the searches, recommendations, prices, and promotions (Montgomery and Smith 2009). AI can automate service encounters and give a personalized and relevant information on a variety of devices (Bock et al. 2020).

2.3.2.4 Order Billing/Payment Management

AI can help e-commerce companies in three key areas: invoicing, payment optimization, and fraud detection (Mejia 2019). Billing and invoice processing are the most burdensome and complicated tasks in business. However, AI applications can help businesses in matching customer invoices with received payments (Dwivedi et al. 2019). Manual and semi-automated billing processes cannot handle a huge number of customer payments, but AI-enabled billing systems support features like invoice segregation, data extraction, invoice generation, etc. and can handle a huge amount of data to avoid any anomalies, inconsistencies, and disparities within the invoices (Bajpai 2020). Online transaction is one of the most convenient ways to perform payment with the help of computer and internet. Customers can use their e-wallets, credit card, debit card, and online banking credentials for an online transaction (Kalia 2016; Kalia et al. 2017a). Despite various security measures, a risk in terms of fraud is associated with every online transaction (Papadopoulos and Brooks 2011). AI-powered billing software can prevent fraud from occurring in the first place by activating the automatic decision-oriented and sophisticated fraud detection system (Khattri and Singh 2018). For example, Fraugster uses the historical data related to transaction, billing, shipping addresses, and IP connection to detect the payment fraud (Canhoto and Clear 2020).

2.3.3 Service and Support

2.3.3.1 Order Scheduling/Fulfillment Delivery

Order scheduling and fulfillment include the tasks related to pick-up or delivery of products and services at the right place and time in the right quantity and quality (Weber and Schütte 2019). Due to irregular order patterns, limited time for order processing, and short-term delivery schedules, e-commerce industry requires extremely efficient fulfillment processes (Leung et al. 2018). Here, the AI systems can actively monitor and optimize these processes by considering order demand factors and product characteristics to automate the perfect logistics strategy (Lam et al. 2015; Paschen et al. 2019). Researchers argue that sustainable supply chains and reverse logistics are the predominant themes for the present as well as future researches in AI (Dhamija and Bag 2019). Therefore, leading e-commerce companies like Amazon.com are investing in robotics and space-age fulfillment technologies (drones for order delivery) (Dirican 2015). Apart from the fulfillment, reverse logistics is another challenge. Products are generally returned without their original packing plus seasonally changing collections, and the product similarity can complicate the process. AI-powered automatic image recognition can compare these returns with catalogue images to sort the products (Kumar et al. 2014).

2.3.3.2 Customer Service and Support

In e-commerce industry, service quality can be a game changer (Kalia et al. 2016, 2017b; Kalia 2017a, b). AI can play a leading role in customer service and support by enhancing satisfaction, improving relationships, personalizing support, and providing recovery in case of service failures. Researchers call it "service AI," and define it as "configuration of technology to provide value in the internal and external service environments through flexible adaptation enabled by sensing, learning, decision-making and actions" (Bock et al. 2020). Therefore, service AI is not just about applying preprogramed decisions but it has a learning ability as well (Makridakis 2017). AI can affect the customer satisfaction because AI-based service is more reliable, of high quality, consistent, continuously available (24/7), and less susceptible to human errors arising due to fatigue and bounded rationality (Huang and Rust 2018). Similarly, it is easier for e-commerce companies to undertake marketing activities to establish, develop, and maintain relationships with customers (Lo and Campos 2018). For instance, virtual assistants can notify millions of users or analyze their purchases, returns, or loyalty card information and provide services which are beyond the human ability. AI-enabled systems can create comprehensive profiles of current or potential customers by using structured and unstructured data related to their psychographic, demographic, and webographic characteristics, and online buying behavior (frequency, recency, type, and size of past purchases), and process it with the help of machine learning and predictive algorithms to strengthen the customer relationship efforts and prospecting of potential customers (Lo and Campos 2018; Paschen et al. 2019). AI can provide a high-quality personalized experience to customers by learning to speak in multiple languages, identify customers' emotional states, or retrieve information for them. A dissatisfied customer can spread a negative word of mouth. However, firms can mine the high-quality customer feedbacks to develop service recovery strategies (Lo and Campos 2018).

2.4 CONCLUDING REMARKS

AI holds an elaborate and key role in various segments/processes of e-commerce business enterprises. To achieve this objective, the role of AI under e-commerce business processes like market research, market stimulation, terms negotiation, order receipt, order selection and priority, order billing/payment management, order scheduling/fulfillment delivery, and customer service and support has been detailed. We noticed that the possible applications of AI for marketing, transaction processing, and service and support for e-commerce business are numerous. At the pace of the current technological developments, AI will graduate from merely a tool for data and information processing (weak AI) to an independent system capable of making human decisions (strong AI).

REFERENCES

Aggarwal A (2018) Resurgence of AI during 1983–2010. https://www.kdnuggets.com/2018/02/resurgence-ai-1983-2010.html. Accessed 16 May 2020.

American Marketing Association (2017) Definitions of marketing. https://www.ama.org/the-definition-of-marketing-what-is-marketing/. Accessed 17 May 2020.

Bajpai K (2020) Artificial intelligence in billing and invoice processing: The future is here! https://www.elorus.com/blog/artificial-intelligence-billing-invoice-processing/. Accessed 16 May 2020.

Berthon PR, Pitt LF (2018) Brands, truthiness and post-fact: Managing brands in a post-rational world. *J Macromarketing* 38:218–227.

Bock DE, Wolter JS, Ferrell OC (2020) Artificial intelligence: Disrupting what we know about services. *J Serv Mark*. doi: 10.1108/JSM-01-2019-0047.

Campbell C, Sands S, Ferraro C, et al. (2020) From data to action: How marketers can leverage AI. *Bus Horiz* 63:227–243. doi: 10.1016/j.bushor.2019.12.002.

Canhoto AI, Clear F (2020) Artificial intelligence and machine learning as business tools: A framework for diagnosing value destruction potential. *Bus Horiz* 63:183–193 doi: 10.1016/j.bushor.2019.11.003.

Dawid H (1999) Bounded rationality and artificial intelligence. In: *Adaptive Learning by Genetic Algorithms*. Springer-Verlag, Berlin, Heidelberg. doi: 10.1007/978-3-642-18142-9_2.

Dhamija P, Bag S (2019) Role of artificial intelligence in operations environment: A review and bibliometric analysis. *TQM J*. doi: 10.1108/TQM-10-2019-0243.

Dirican C (2015) The impacts of robotics, artificial intelligence on business and economics. *Procedia Soc Behav Sci* 195:564–573. doi: 10.1016/j.sbspro.2015.06.134.

Dumitriu D, Popescu MA (2020) Artificial Intelligence solutions for digital marketing. *Procedia Manuf* 46:630–636 doi: 10.1016/j.promfg.2020.03.090.

Dwivedi YK, Hughes L, Ismagilova E, et al. (2019) Artificial intelligence (AI): Multidisciplinary perspectives on emerging challenges, opportunities, and agenda for research, practice and policy. *Int J Inf Manage* 101994. doi: 10.1016/j.ijinfomgt.2019.08.002.

Florez-lopez R, Ramon-jeronimo JM (2009) Marketing segmentation through machine learning models: An approach based on customer relationship management and customer profitability accounting. *Soc Sci Comput Rev* 27:96–117.

Frey CB, Osborne MA (2017) The future of employment: How susceptible are jobs to computerisation? *Technol Forecast Soc Change* 114:254–280. doi: 10.1016/j.techfore.2016.08.019.

Holmqvist J, Van Vaerenbergh Y, Grönroos C (2017). Language use in services: Recent advances and directions for future research. *J Bus Res* 72:114–118. doi: 10.1016/j.jbusres.2016.10.005.

Hruschka H, Natter M (1999) Comparing performance of feedforward neural nets and K-means for cluster-based market segmentation. *Eur J Oper Res* 114:346–353.

Huang M, Rust RT (2018) Artificial intelligence in service. *J Serv Res* 21:155–172. doi: 10.1177/1094670517752459.

Huang S, Lin F (2005) Designing intelligent sales-agent for online selling. In: *Proceedings of the 7th International Conference on Electronic Commerce.* Association for Computing Machinery, New York, NY/Xi'an, China, pp. 279–286.

IBM (2020) Getting started with natural language understanding. https://cloud.ibm.com/docs/services/natural-language-understanding?topic=natural-language-understanding-getting-started#analyze-phrase. Accessed 17 May 2020.

Jarek K, Kozminskiego AL, Mazurek G (2019) Marketing and artificial intelligence. *Cent Eur Bus Rev* 8:46–55 doi: 10.18267/j.cebr.213.

Järvinen J, Taiminen H (2016) Harnessing marketing automation for B2B content marketing. *Ind Mark Manag* 54:164–175. doi: 10.1016/j.indmarman.2015.07.002.

Jusoh S (2018) Intelligent conversational agent for online sales. In: *10th International Conference on Electronics, Computers and Artificial Intelligence (ECAI).* IEEE, Iasi, Romania, pp. 1–4.

Kahneman D, Tversky A (1979) Prospect theory: An analysis of decision under risk. *Econometrica* 47:263–292.

Kalia P (2016) Tsunamic e-commerce in India: The third wave. *Glob Anal* 5:47–49.

Kalia P (2017a) Service quality scales in online retail: Methodological issues. *Int J Oper Prod Manag* 37:630–663. doi: 10.1108/IJOPM-03-2015-0133.

Kalia P (2017b) Webographics and perceived service quality: An Indian e-retail context. *Int J Serv Econ Manag* 8:152–168. doi: 10.1504/IJSEM.2017.10012733.

Kalia P, Arora R, Kumalo S (2016) E-service quality, consumer satisfaction and future purchase intentions in e-retail. *e-Service J* 10:24–41. doi: 10.2979/eservicej.10.1.02.

Kalia P, Kaur N, Singh T (2017a) E-commerce in India: Evolution and revolution of online retail. In: Khosrow-Pour M (ed) *Mobile Commerce: Concepts, Methodologies, Tools, and Applications.* IGI Global, Hershey, PA, pp. 736–758.

Kalia P, Law P, Arora R (2017b) Determining impact of demographics on perceived service quality in online retail. In: Khosrow-Pour M (ed) *Encyclopedia of Information Science and Technology*, 4th edn. IGI Global, Hershey, PA, pp. 2882–2896.

Kaplan A, Haenlein M (2019) Siri, Siri, in my hand: Who's the fairest in the land? On the interpretations, illustrations, and implications of artificial intelligence. *Bus Horiz* 62:15–25. doi: 10.1016/j.bushor.2018.08.004.

Kawa A, Pieranski B, Zdrenka W (2018) Dynamic configuration of same-day delivery in e-commerce. In: Sieminski A, Kozierkiewicz A, Nunez M, Ha QT (eds) *Modern Approaches for Intelligent Information and Database Systems.* Springer, Berlin, pp. 305–315.

Kephart JO, Hanson JE, Greenwald AR (2000) Dynamic pricing by software agents. *Comput Netw* 32:731–752.

Khattri V, Singh DK (2018) Parameters of automated fraud detection techniques during online transactions. *J Financ Crime* 25:702–720. doi: 10.1108/JFC-03-2017-0024.

Krasadakis G (2017) AI buyer/seller negotiation agents. https://medium.com/innovation-machine/artificial-intelligence-negotiation-agents-49d666cd9952. Accessed 16 May 2020.

Kumar DT, Soleimani H, Kannan G (2014) Forecasting return products in an integrated forward/reverse supply chain utilizing an ANFIS. *Int J Appl Math Comput Sci* 24:669–682. doi: 10.2478/amcs-2014-0049.

Lam HY, Choy KL, Ho GTS, et al. (2015) A knowledge-based logistics operations planning system for mitigating risk in warehouse order fulfillment. *Int J Prod Econ* 170:763–779. doi: 10.1016/j.ijpe.2015.01.005.

Legg S, Hutter M (2006) A collection of definitions of intelligence. In: *Advances in Artificial General Intelligence: Concepts, Architectures and Algorithms*. IOS Press, Amsterdam, pp. 17–24.

Leung KH, Choy KL, Siu PKY, et al. (2018) A B2C e-commerce intelligent system for re-engineering the e-order fulfilment process. *Expert Syst Appl* 91:386–401. doi: 10.1016/j.eswa.2017.09.026.

Lo FY, Campos N (2018) Blending Internet-of-Things (IoT) solutions into relationship marketing strategies. *Technol Forecast Soc Change* 137:10–18. doi: 10.1016/j.techfore.2018.09.029.

Makridakis S (2017) The forthcoming artificial intelligence (AI) revolution: Its impact on society and firms. *Futures* 90:46–60. doi: 10.1016/j.futures.2017.03.006.

McCarthy EJ (1960) *Basic Marketing: A Managerial Approach*. Richard D. Irvin, Inc., Homewod, IL.

McCarthy J (2007) What is artificial intelligence? https://www-formal.stanford.edu/jmc/whatisai/node1.html. Accessed 17 May 2020.

Mckendrick J (2019) Now, AI can give you an edge … in negotiations. https://www.rtinsights.com/now-ai-can-give-you-an-edge-in-negotiations. Accessed 16 May 2020.

Mejia N (2019) Artificial intelligence in payment processing – Current applications. https://emerj.com/ai-sector-overviews/artificial-intelligence-in-payment-processing-current-applications/. Accessed 16 May 2020.

Merriam-Webster.com (2020) Definition of intelligence. https://www.merriam-webster.com/dictionary/intelligence. Accessed 17 May 2020.

Montgomery AL, Smith MD (2009) Prospects for personalization on the Internet. *J Interact Mark* 23:130–137. doi: 10.1016/j.intmar.2009.02.001.

O'Brien JA, Marakas GM (2011) *Management Information System*, 10th edn. McGraw-Hill/Irwin, New York, NY.

Ostrom AL, Fotheringham D, Bitner MJ (2018) Customer acceptance of AI in service encounters: Understanding antecedents and consequences. In: Maglio PP, Kieliszewski CA, Spohrer JC, et al. (eds) *Handbook of Service Science*. Springer Nature, Cham, pp. 77–103.

Papadopoulos A, Brooks G (2011) The investigation of credit card fraud in Cyprus: reviewing police "effectiveness." *J Financ Crime* 18:222–234. doi: 10.1108/13590791111147442.

Paschen J, Kietzmann J, Kietzmann TC (2019) Artificial intelligence (AI) and its implications for market knowledge in B2B marketing. *J Bus Ind Mark* 34:1410–1419. doi: 10.1108/JBIM-10-2018-0295.

Reynolds M (2017) Chatbots learn how to negotiate and drive a hard bargain. *NewScientist* 234:7. doi: 10.1016/S0262-4079(17)31142-9.

Shankar V (2018) How artificial intelligence (AI) is reshaping retailing. *J Retail* 94:vi–xi. doi: 10.1016/S0022-4359(18)30076-9.

Simon JP (2019) Artificial intelligence: Scope, players, markets and geography. *Digit Policy Regul Gov* 21:208–237. doi: 10.1108/DPRG-08-2018-0039.

Sonia N, Sharma EK, Singh N, Kapoor A (2020) Artificial intelligence in market business: From research and innovation to deployment market. *Procedia Comput Sci* 167:2200–2210. doi: 10.1016/j.procs.2020.03.272.

Stone M, Aravopoulou E, Ekinci Y, et al. (2020) Artificial intelligence (AI) in strategic marketing decision- making: A research agenda. *Bottom Line.* doi: 10.1108/BL-03-2020-0022.

Syam N, Sharma A (2018) Waiting for a sales renaissance in the fourth industrial revolution: Machine learning and artificial intelligence in sales research and practice. *Ind Mark Manag.* 69:135–146. doi: 10.1016/j.indmarman.2017.12.019.

Tegmark M (2017) *Life 3.0: Being Human in the Age of Artificial Intelligence.* Alfred A. Knopf, New York, NY.

Weber FD, Schütte R (2019) State-of-the-art and adoption of artificial intelligence in retailing. *Digit Policy Regul Gov* 21:264–279. doi: 10.1108/DPRG-09-2018-0050.

Wilson RD, Bettis-outland H (2020) Can artificial neural network models be used to improve the analysis of B2B marketing research data? *J Bus Ind Mark* 35:495–507. doi: 10.1108/JBIM-01-2019-0060.

Zanker M, Rook L, Jannach D (2019) Measuring the impact of online personalisation: Past, present and future. *J Hum Comput Stud* 131:160–168. doi: 10.1016/j.ijhcs.2019.06.006.

3 ABC of Digital Era with Special Reference to Banking Sector

Shaveta Chugh
Khalsa College for Women

CONTENTS

3.1 INTRODUCTION

Artificial intelligence (AI), a branch of computer science, is a rapidly developing technology across the globe. An interdisciplinary science[1] with multiple approaches is concerned with building and edifying smart technology, for performing a variety of tasks[2]. With the passage of time, a continuous improvement in technology allows the various sectors of the society to adopt artificial technology. It is a fact that India is considered as a tech-hub but among the various sectors, the banking sector is one of the initial adopters of AI. Banks executed the implementation of technology in a variety of ways. AI, in the banking sector, is not only helpful in retaining the clients but also assists in improving the course of action and personalizing the consumer's experience[3]. It is the means to alter the maximum critical customer facing tasks and helps in maintaining the competitive edge.

3.2 ARTIFICIAL INTELLIGENCE IN BANKING SECTOR

In Indian banking sector, technology is considered as the foundation stone. In the mid- to late 1990s, Indian banking sector started using the technology to embark its journey. The banking sector then starts deploying the latest technologies starting from its routine transactions to core banking operations. The adoption of technology as AI, a larger digital wave, at such a vast pace is considered as the fourth Industrial Revolution.

AI is nothing but the development of computer systems at the workplace, which help the various sectors in performing the task naturally that requires human intelligence like voice recognition translation between languages etc. Quite aggressively,[4] Indian banking sector is using AI in the current scenario. As per the report given by PwC FinTech (India) 2017,[5] the global investment in the applications of AI was US $ 4 billion in 2015,[6] and in 2017, it touched $ 5.1 billion.[7] The results were further seen in the next financial year 2018, which shows that India's digital lending stood at US $ 75 billion,[8] and it was further expected by the end of FY 2023, the lending will rise to US $ 1 trillion,[9] which includes five times increase in the digital expenditures.

According to the information provided by the multinational IT consultancy firm Gartner, Inc., having its headquarters in the United States, in the financial year 2020 the Indian banking sector continues its pace while investing in digital business and shows an increase in investment by 9.1% as compared to FY 2019.[10] In FY 2020, a total investment of $ 11 billion was expected.[9]

3.3 ABC OF DIGITAL ERA IN BANKING SECTOR

As the banking sector is moving toward the adoption of digitalization, the most frequently asked question is, "*How do they become Digital Banking superstars?*" We are approaching that world where the future of finance is ever increasing where the banks provide a financial service to customers, having their roots in e-commerce. Banks will have to work hard from just providing basic services[11] and have to look forward in order to become the bank of the future. This can be possible only by changing the definition of ABC in the respective sector, where:

A—Artificial Intelligence
B—Big Tech
C—Core Banking and Cloud

3.3.1 A AS ARTIFICIAL INTELLIGENCE

Banks are adopting the AI technologies[12] in their sector in the form of automation, chatbots[13], analytics etc. The adoption of these facilities will make the bank smarter, faster, and more robust. Beyond the tech sector, while taking in view of investment in AI, the banking sector is considered as one of the biggest spenders in that field. Table 3.1 shows expenditure by various sectors in AI.

TABLE 3.1
Expenditure by Various Sectors in Artificial Intelligence

$ Millions	2016	2019
Telecommunications	50	300
Education	100	450
Utilities	100	500
Insurance	200	650
Transportation	200	700
Banking	1900	7500
Retail	1100	5000

Source: https://www.citibank.com/commercialbank/insights/assets/docs/2018/The-Bank-of-the-Future/10/.

3.3.2 B as Big Tech

While taking tension of the future, the banks will not ignore the challenges being faced by them by Big Tech. With the entry of Big Tech in the market, the situation changes. Traditional banking nowadays is becoming an object of the past. Big Tech has its own pre-owned extent and client reach, and it is quite possible that it can be greater than banks. The companies like Amazon[14], Facebook etc. have captured the market with ever escalating share of consumer's time and attention. These sites consider the financial system as a tool to grab more customers, and when they have entered into the zone of payment and finance, they have gone gigantic. The Big Techs recreated the form of making payment with the creation of mobile platforms, which is considered as a favorable policy not only by the middle-class people but by the government also. FinTechs were considered as the service providers to the banks but Big Techs are more friendly[15]. But in order to survive in this digital hi-tech era, the banking sector must adopt a data-driven and customer-centric mindset. Figure 3.1 shows the approximate number of users for mobile wallet providers in India.

A recently launched business app, WhatsApp, helps the small business houses and firms to open a different account having messaging tools, and the results can be shown from the image. Indian banking sector, while considering Big Techs, has taken a step to enter in the world of digital transactions, which include NEFT/RTGS, debit and credit cards payment, BHIM UPI etc. This helps the banking sector in increasing their electronic payment transaction by 95.4% in 2018–2019, up from 92.6%[16] in the previous year.

3.3.3 C as Core Banking and Cloud

During the 1950–1960s, banks invested in mainframe computers, which was the first investment made by banks in technology.[17] Over more than 50 years, some

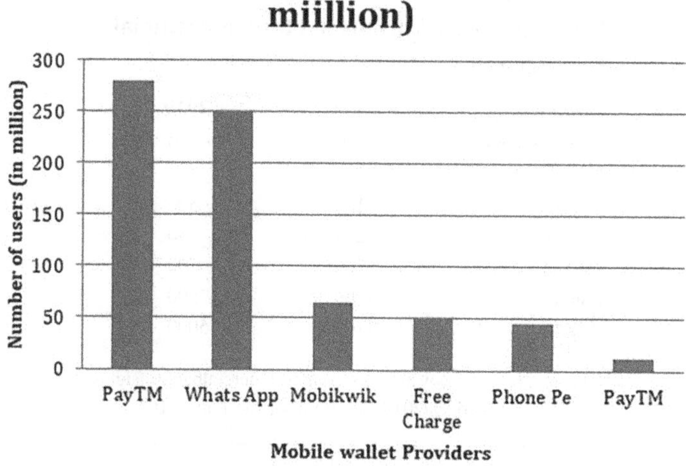

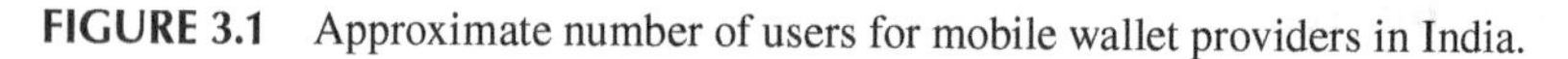

FIGURE 3.1 Approximate number of users for mobile wallet providers in India.

(Source: https://studfile.net/preview/9106013/page:13/.)

banks are still using legacy systems and are working on outdated software packages. Cost and time constraints restrict the sector in completely shifting to the new world of digitalization. During the past few years, the refurbishment of core infrastructure took place because the outdated banking IT systems have reached the point of redundancy. Few factors that contribute to a drive towards IT investments include:

- New and improved technology
- More focus on data quality, accessibility etc.
- Rise in the expectations of customers
- More focus on cost-saving

Cloud and modern technology help the banking sector in improving their efficiency, swiftness, and speed to market. No doubt they may provide better marketing opportunities but, in the coming future, the IT cost of banks is going to be even higher. Table 3.2 shows the percentage allocation of the annual budget.

3.4 OPPORTUNITIES AND CHALLENGES IN BANKING SECTOR DUE TO DIGITALIZATION

3.4.1 Opportunities

The banking sector in India has started deploying AI in the areas of business management and financial operations. Some of the areas where AI is used in banking industry include the following:

TABLE 3.2
Allocation of Percentage of Annual Budget in Various Segments

Banking	8.7
High	5.2
Tech	4.2
Utilities	3.4
Insurance	3.4
Healthcare	3.4
Financial services	3.1
Agriculture	1.7
Energy	1.1

Source: https://www.citibank.com/commercialbank/insights/assets/docs/2018/The-Bank-of-the-Future/10/.

1. **Customer Interface**: Banks in the present era are using a virtual customer support like chatbot, mobile applications etc. to assist them in resolving their queries.
2. **Personalized Financial Services**: Banks provide personalized service to the customers, which includes financial advisors and planners that help them in taking appropriate financial decisions at the right time. The advisors analyze the market, give recommendations to the customers, and help them in their financial planning.
3. **Credit Scoring and Loan Decisions**: Banks use AI to calculate the credit scores on the basis of which the firms can make rapid decisions related to credit by analyzing the spending habits, family history, and the banking transactions of individuals.
4. **Digital Wallets**: Digital wallets are considered as the future of real-time disbursement technologies. They include Amazon pay, PayPal etc. They help the customers in decreasing their dependencies on liquid cash.[4] They introduce the concept of anytime payment, hence increasing the reach of money to new heights.
5. **Fraud Detection and Risk Management**: The technology by continuously monitoring the transactions helps in the detection of various kinds of frauds and malpractices, and ultimately helps in the detection of potential risk at an appropriate time.
6. **Voice-Assisted Banking**: With the advancement of AI, the physical presence of customers is fading down. They have started using the services like voice commands, touch screens etc., rather than interacting with the employees of banks. The robotics[12] like machine-like ATM help them in resolving their queries and providing appropriate solutions to their problems by connecting the customers to various banking services. This will lead to a reduction in human error[18] and ultimately an increase in their efficiency.

3.4.2 Challenges

1. **Lack of Skilled Personnel**: The banking staff is not tech-savvy. A limited number of employees in the sector can use AI effectively. In order to handle this challenge, a proper training should be given to the employees.
2. **Data-Driven**: AI technologies are data hungry. They require enormous amounts of data for providing better results. So, another problem confronted by the sector is the lack of usable data that can be investigated by AI, as the adage "Garbage In and Garbage Out" still applies to banks.
3. **Expensive**: The fact is that there are some problems which can be solved with the help of traditional intelligence tools and will not get much benefit with the adoption of AI. So, in those cases, the adoption of AI is an expensive affair.
4. **Continual Technology Changes**: In every sector, the choice of technology plays an important role. Banking sector has started adopting the technologies available in the market. With the significant changes in the market, the technology adopted by the banks will become obsolete with the introduction of new tools and equipment. This will lead to an increase in expenditure of the banks.

3.5 ARTIFICIAL INTELLIGENCE USED BY FOUR BIG BANKS OF INDIA

The leading commercial banks, using AI, in India include SBI, HDFC bank, ICICI bank, and Axis bank. The AI used by these banks, and its purpose and benefits are well explained in Tables 3.3-3.6.

3.5.1 State Bank of India

SBI is India's largest public sector bank and is ranked 236th[19] in the list of the world's biggest corporation of 2019, i.e., Fortune Global 500. It uses AI and launched a national hackathon with a name, *Code For Bank*.[20,21] It was basically for developers

TABLE 3.3
Artificial Intelligence Used by State Bank of India

State Bank of India	Chapdex	An AI that includes checking of cameras that set up in the branch of the bank and capturing the outer appearances of the clients.[18] It is just a technology to analyze the facial expressions of clients
	SIA Chat Assistant	An AI technology, developed by Payjo, included a voice assistant that will help in handling the queries of customers. It is capable of handling nearly 10,000 enquiries per second or 864 million queries in a day. Such a useful technology handles approximately 25% of queries being processed by Google daily[22]

TABLE 3.4
AI Used by HDFC Bank

HDFC Bank	Electronic Virtual Assistance (EVA)	It helps in integrating the knowledge from thousands of sources and provides response in less than 0.4 seconds.[23] It saves the ample amount of time and resources of bank employees. It evacuates the need to search, persuade, and call
	Chatbot	It is a bank's Facebook messenger that helps the customers in tracking the stock prices before investing in the stock market. It helps an investor in creating their own portfolios, buying and selling stock on the basis of chat, getting the latest trends about the market, and receiving recommendations on the basis of which transactions can take place

TABLE 3.5
Technology by ICICI Bank

ICICI Bank	iPal	Services provided by iPal are broadly categorized into the three main categories: 1. It involves simple yet valuable FAQs which a customer wants to ask his bank executive or employee. It comprises those questions that have simple structured responses 2. It includes the involvement of transactions in the form of bill payment, transfer of amount from one account to another, recharge of mobiles, and other utilities 3. It includes guiding the customers with basic features like guiding them in changing the PIN of their ATM card This feature has connected about 3.1 million customers and gave replies to approximately 6 million queries with 90% accuracy[23]

and students, which motivates them to turn up with innovative, thought-provoking ideas which include latest technologies and must be beneficial for the banking sector. The AI used in State Bank of India is shown in Table 3.3.

3.5.2 HDFC Bank

HDFC is India's leading private sector bank that offers a variety of services to its customers. It is considered as India's largest bank by market capitalization as of March 2020.[19] The AI used by HDFC bank is shown in Table 3.4.

TABLE 3.6
Technology Adopted by AXIS Bank

AXIS Bank	AI and natural language processing (NLP)-enabled app	This app helps the customers in providing them guidance regarding their financial and nonfinancial transactions. It also helps them in resolving their queries by giving a response to their FAQs[24]. Presently, it is available on Facebook and the website of the bank, but soon the service will be extended to the mobiles of the customers This technology will help the banks in reducing the turnaround time (TAT)—as saving bank account opening has now reduced by 90%, current accounts by 92% etc.[16]
	AXIS Aha	It is an intelligent working banking assistant,[25] which will address the queries on a daily basis and facilitate on-the-go and instant digital support for the customers of the bank. The online transactions like mobile recharge and payment for utilities etc. can be done on a single click

3.5.3 ICICI Bank

ICICI Bank ltd. is India's second largest private sector bank that deals in providing a vast range of banking products and services to its retail and corporate customers. The technology adopted by the ICICI bank is shown in Table 3.5.

3.5.4 AXIS Bank

AXIS Bank is India's third largest private sector bank. It includes providing services to large and mid-corporate, MSME, agriculture, and retail businesses. The technology opted by AXIS bank is shown in Table 3.6.

3.6 CONCLUSION

After demonetization, the trend of digitalization is taking place in almost all sectors of the society, especially banking. In this digitalized era, the traditional banking has changed and has started adopting new technologies like AI, Cloud etc. in order to increase their efficiency and cut their operating cost. Banking sector is in its transition phase, i.e., shifting from the budding stage to the era of AI. AI brings a change in the business processes and customer facing services in the respective sector. No doubt, it would lead to relaxing the task of banks as well as accommodating the customers but, on the other hand, the complications like an increase in cybersecurity threat have taken place. So, the main duty of the banks is to adopt technologies like block chain that help them in creating active defense mechanisms against cybercrimes.

REFERENCES

1. Luber, S. (2011, March). Cognitive science artificial intelligence: Simulating the human mind to achieve goals. In *2011 3rd International Conference on Computer Research and Development* (Vol. 1, pp. 207–210). IEEE, Shanghai, China.
2. Yadav, R., & Yadav, R. (2018). Review on artificial intelligence: A boon or bane to humans. *International Journal of Scientific Research in Computer Science, Engineering and Information Technology*, 3(4). ISSN: 2456-3307.
3. Sabharwal, M. (2014). The use of artificial intelligence (AI) based technological applications by Indian Banks. *International Journal of Artificial Intelligence and Agent Technology*, 2, 1–5.
4. Jewandah, S. (2018). How artificial intelligence is changing the banking sector – A case study of top four commercial Indian banks. *International Journal of Management, Technology and Engineering*, 8(7), 525–530.
5. Fintech Trends Report India (2017). https://www.pwc.in/assets/pdfs/publications/2017/fintech-india-report-2017.pdf.
6. Salunkhe R. (2019). Role of artificial intelligence in providing customer services with special reference to SBI and HDFC Bank. *International Journal of Recent Technology and Engineering (IJRTE)*, 8(4). ISSN: 2277-3878.
7. Sindhu, J., & Namratha, R. (2019). Impact of artificial intelligence in chosen Indian Commercial Bank – A cost benefit analysis. *Asian Journal of Management*, 10(4), 377–384.
8. Parveen, N. (2020). Analytical study on price movement of particular banking stocks using relative strength index (RSI). Purakala with ISSN 0971-2143 is an *UGC CARE Journal*, 319
9. Digital Transactions: Digital payments in India to reach 1 trillion by 2023, https://cio.economictimes.indiatimes.com/news/enterprise-services-and-applications/digital-payments-in-india-to-reach-1-trillion-by-2023-credit-suisse/62940348.
10. IT spend in banking and securities sector in India to grow 9% in 2020: Gartner. https://economictimes.indiatimes.com/industry/banking/finance/banking/it-spend-in-banking-and-securities-sector-in-india-to-grow-9-in-2020-gartner/articleshow/71702670.cms?utm_source=contentofinterest&utm_medium=text&utm_campaign=cppst.
11. Joseph, D. T. (2020). Banking Industry – The vital cog in the wheel of artificial intelligence. *Studies in Indian Place Names*, 40(16), 203–210.
12. Dhanabalan, T., Sathish, A. (2018). Transforming Indian industries through artificial intelligence and robotics in industry 4.0. *International Journal of Mechanical Engineering and Technology*, 9(10), 835–845.
13. Agarwal, B., Aggaarwal, H., Talib, P. (2019). Application of artificial intelligence for successful strategy implementation in India's banking sector. *International Journal of Advanced Research*, 7(11), 157–166.
14. India on the frontline of Digital Finance. https://studfile.net/preview/9106013/page:13/.
15. Artificial Intelligence in Fintech Market: Groth, trends and forecast. https://www.dqindia.com/artificial-intelligence-fintech-market-growth-trends-forecast/.
16. Annual Report RBI. https://m.rbi.org.in/Scripts/AnnualReportPublications.aspx?Id=1264.
17. The bank of the future. https://www.citibank.com/commercialbank/insights/assets/docs/2018/The-Bank-of-the-Future/10/.
18. Vijai, D. C. (2019). Artificial intelligence in Indian Banking Sector: Challenges and opportunities. *International Journal of Advanced Research*, 7(5), 1581–1587.
19. "Largest bank by market capitalization". *Money Control*, https://www.moneycontrol.com/stocks/marketinfo/marketcap/bse/banks-private-sector.html.

20. Lakshminarayana. N, Deepthi B. R. (2019). Advent of artificial intelligence and its impact on top leading commercial banks in India – Case study. *International Journal of Trend in Scientific Research and Development (IJTSRD)*, 3(4), 614–616.
21. AI applications in top 4 Indian bankshttps://emerj.com/ai-sector-overviews/ai-applications-in-the-top-4-indian-banks/.
22. Payjo launches SBI intelligent assistant chatbot with AI capabilities. https://www.digit.in/press-release/machine-learning-and-ai/payjo-launches-sbi-intelligent-assistant-sia-chatbot-with-ai-capabilities-37256.html.
23. Kumari, R., Kochikunnel, N. J. (2018). Application of Fintech in banking sector with reference to artificial intelligence. *International Journal of Social Science and Economic Research*, 3(9). ISSN: 2455-8834.
24. Introducing AXIS-AHA An AI- led virtual assistant to enhance online customer experience. https://www.axisbank.com/docs/default-source/press-releases/introducing-axis-aha!-an-ai-led-virtual-assistant-to-enhance-online-customer-experience.pdf?sfvrsn=6b0fb355_0
25. AXIS AHA, Overview; https://www.axisbank.com/bank-smart/axis-aha/overview.

4 Artificial Intelligence in Predictive Analysis of Insurance and Banking

L. R. Paul, L. Sadath, and A. Madana
Amity University

CONTENTS

4.1 INTRODUCTION

Artificial intelligence (AI) is the study in which computer systems are vested with the ability to carry out tasks, which usually depends on human intelligence. AI allows an advanced technology to replace humans in the processes of data collection and analyses, and even in the process of final decision-making [1].

The use of modern-day technologies has increased drastically in the banking sector. Hence, AI is now implemented in many banks around the world. Earlier banks required a human involvement for their functioning. Now with the implementation of various technologies in banks, it resulted in a maximum performance. Hence, one can find AI in banks in services related to personalized finances where AI helps in examining stock markets and offers recommendations as per the monetarist aims of customers. Analyzing massive amounts of data, and carrying out various calculations and prediction capabilities helped in automated loaning decisions to be made by AI for achieving better returns. Unique ability of AI in language processing helped to advance effectiveness of the processes in banks, which thereby helped in reducing human errors [2]. The power of speech processing with AI helped to automate various customer service requests, and this helped to reduce the waiting time of customers waiting for a customer representative to attend their calls [3].

Customers of insurance and banking industry now expect quick efficient services. Millennials are all about convenience. In insurance, areas such as processing claims, detecting frauds, underwriting, customer service etc. have evolved to involve technology enhancing these processes. The major reason behind this proliferation of AI is better data analysis and insight.

Banking and financial services industry also finds several applications of AI such as lending advisors, bond portfolio planning and gaining trading insights, fault diagnosis etc. Calculations such as rates of return, time-series analysis, actualization indexes etc. are specific to the domain of banking and finance, and they are carried out through computational techniques of risk ranking, linear programming algorithms, multiattribute scoring, protocol verification etc. [4].

Investment banks adopt AI to prevent fraudulent activities and to provide standard and improved customer service and real-time solutions, for digital documentation etc. Banks use machine intelligence and Big Data to gather customer information in order to evaluate their credit worthiness and offer personalized services. AI algorithms make it possible to develop highly refined investment strategies and curb money laundering by analyzing internal data, which is available publicly within the customer network. AI allows banks to asses risk using large volumes of data, to understand and adapt to the evolving financial landscape, to be more consistent in their operations, to add value, and to make quality decisions [5].

AI is also used in the insurance sector. A consumer normally purchases an insurance policy and makes claims, and this is the basic connection between a consumer and the insurer. Currently, AI in the insurance sector is used to reinforce the capacities and knowledge of the insurer and not the consumer [6]. If the insurer has more knowledge of the past claims of the customer and if there is very less or no claims at all, then the insurance product can be sold for a cheaper price, which can be competitive in the market.

This chapter discusses some of the key applications of AI in insurance, banking, and finance, such as predictive analysis, genetic algorithms (GAs), and anomaly detection.

4.2 PREDICTIVE ANALYSIS AND ITS APPLICATIONS

Data mining and predictive analysis have their widespread use in business and are known by different names such as knowledge discovery, sense making etc. However, the most recent name given to them is data science. Data mining is making use of enormous data sets to identify trends or patterns to solve problems. It involves the usage of automated methods to systematically analyze huge chunks of data, and draw reasonable and sensible conclusions. This kind of data analysis helps us to prove or disprove a hypothesis, discover unknown information, identify relationships between data, and classify and anticipate additional data for future events. One of the most common examples of the predictive analysis in e-commerce sector is the emergency response plan of Walmart. Walmart effectively uses historical data regarding the point of sale to increase the customer satisfaction. Major weather events affect the purchasing behavior of customer, such as bulk buying of items like bottled water, pop-tarts, duct tape etc. before a storm. The bottled water and duct tape were bought because they were part of the recommendations given by the government, so this is a confirmation; however, increase in the sales of pop-tarts is surprising and so it is the discovery of a new relationship. This knowledge of customer buying behavior can help Walmart meet their demands by keeping their supply ready. Anticipating such needs of customers allowed Walmart to respond timely during Hurricane Katrina. Predictive analysis is an extension to the data mining process; it can be considered as a complement to it. It involves the creation of models used to anticipate future events by making use of past trends or patterns. Data mining and predictive analysis are just as much as analytical as machine learning and math. The analysis process is 80% preparation of data and 20% analysis [7].

The tools of data science, predictive analysis, and Big Data will significantly influence the way supply chains are managed, creating new challenges as well as opportunities. The new data-rich environment has even made certain standard practices in SCM (supply chain management) obsolete.

Predictive analysis is a subcategory of data science. It is associated with several quantitative and qualitative approaches, in contrast to statistics, which is quantitative. Predictive analytics is not just about forecasting; it answers questions as to what would have happened in the past, if conditions were different. In contrast to optimization which involves finding the maximum or minimum function subject to certain constraints, predictive analytics also aims to find out what are the characteristics of a system that is not optimal [8].

Predictive analytics allow companies to optimize their existing processes, gain more insight into customer behavior, identify opportunities, and anticipate challenges and mitigate them. It involves a large set of techniques and methodologies, which consist of mathematics, statistics, and AI. It adds a heavy amount of data management on top of all this. Predictive analytics is inductive in nature. Rather than presuming things, it lets data take the lead and explores that data using neural computing, machine learning, statistical techniques, AI, computational mathematics etc. to identify the relationships or patterns in this data that can manifest itself in the future. Predictive analysis has moved far from just statistics, and today, with the help of advanced computers and technologies, techniques like decision trees, neural networks, support vector machines (SVMs), advanced mathematical algorithms, and GAs are employed, which involve complex calculations and iterations to make sense of the enormous amounts of data.

Supervised and Unsupervised Predictive Analyses: Supervised learning involves using historical data that consists of the results that you are attempting to predict.

It includes techniques such as regression, classification, time-series analysis etc. Unsupervised learning does not use historical data or results in predictive modeling, but uses descriptive statistics and does not predict a target value [9].

A significant application of the predictive analysis is the forecasting of stock prices and their volatility. The relation between return and risk is one of the most important aspects of modern finance. That is exactly why measuring and forecasting volatility is an essential area in asset pricing and risk management. Tools for modeling the dynamics of volatility are also used in other domains of economics as well as social sciences, natural sciences, and even medicine. It is also found useful in agricultural economics. Deregulations in utility sector have also led to several novel applications in volatility modeling of gas and power prices.

Engle (1982) proposed the ARCH (auto-regressive conditional heteroskedasticity) model, which was the key foundation in modeling and forecasting volatility as a time-varying function of the present information. Bollerslev (1986) proposed the GARCH (generalized autoregressive conditional heteroskedasticity) model class, which is infamous for its applications in volatility forecasting. The basic GARCH model assumes that future conditional variances are influenced by symmetrical shocks, while some assets respond asymmetrically to past returns, and it can lead to larger volatilities in the future. T-GARCH, A-GARCH, and exponential GARCH models are commonly used GARCH models to explain this type of asymmetry. These models imply that shocks to the volatility show an exponential decay [10].

Correlation forecasts are another domain of application of predictive analysis. In contrast to volatility, correlations consider the returns of two or more assets. If a portfolio is made up of average correlations, it indicates a symmetry. That is, the performance of the investment will not be very good in down markets as these correlations will increase. Correlations are linked to the business cycle, and the change in the economic situations can influence the expected returns and hence the correlation structure. It was concluded that the correlation is not constant from a study by Longin and Solnik in which they estimated a GARCH model, which produced a measure of conditional monthly correlation, and this was assumed to be constant. The model was useful for testing the correlation constancy, but it is not clear as to how it can be used for correlation forecasting. Correlation forecasting is essential in the valuation of derivative securities [11].

Prices of securities often exhibit large jumps, and using a standard GARCH model might lead to the overestimation of volatility. In contrast, unconditional volatility forecasts show forecasts larger than they actually should be. Similarly in the case of correlation estimates, if only one of the stocks tends to show a large jump in prices, then the estimate is biased towards zero. If the eco-jumps are of the same sign, then it biases the estimates towards (minus) 1 [12].

DCC (dynamic conditional correlation) estimators have the flexibility of univariate GARCH models, and at the same time, they are not as complicated as multivariate GARCH models. They have significant computational advantages over MV-GARCH as the number of parameters that need to be estimated does not depend on the number of series that has to be correlated. DCCs are a generalization of the CCC (constant conditional correlation) proposed by Bollerslev in 1990 [13].

Range-based DCC model combines the CARR (conditional autoregressive range) model with DCC framework, and it is able to outperform the standard DCC model, which is return-based in terms of forecasting ability. This is because range data is more efficient than return data in volatility estimation. This is true in the case of in-sample as well as out-of-sample analysis [14].

ARIMA (autoregressive integrated moving average) is another significant method of forecasting to predict the time-series values based on the data that is input into the model. ARIMA is a robust method for capturing the dynamic behavior of time series in case of short time periods. The changing trends or seasonality component does not adversely affect the effectiveness of this method of forecasting. ARIMA performs significantly better than HoltWinters technique of forecasting as well as forecasting using the aggregate of trend values [15].

A sentiment analysis approach coupled with the predictive analysis is employed to understand the relationship between future stock price movements and stock micro-blog sentiments. Brief yet succinct nature of stock micro-blogs along with their high volume, frequency, and real-time posting makes stock micro-blogging a key driver of events in the stock market and thus contributes to its predictive power in the field. It is also assumed that bearish sentiments have a better predictive accuracy than the bullish sentiment, which is more of wishful thinking, as it gets more attention from investors [16].

Studies also show Twitter feeds as a good source of data for the predictive sentiment analysis. Positive sentiments in the tweets predict a similar upward movement in prices of stocks. Classifying tweets using SVM neutral zone helps us to improve the correlation between tweets and closing price of stock [17].

The trend deviation in wealth has a significant predictive capability in forecasting large portfolios' returns. Without considering multiperiod portfolio decisions, if we focus on predictability in asset returns, it can be a more reliable measure because we may not have to depend on simulation techniques that can considerably reduce the amount of preassigned variables that can be included in the predictive regression. Predictive regression can include several variables such as liquidity, macroeconomic, and lagged returns [18]. Predicting bank failures is another significant application of this. Using computer-based EWS (early warning systems) for making these predictions may have sampling limitations. Studies that employed logit analysis (parametric approach) and trait recognition (nonparametric approach) indicated that both of them performed well in classification results, but the predictive accuracy and stability trait recognition were found to be more efficient. Nonparametric EWS provides an essential information regarding the impending viability of banks, and trait recognition is potentially useful in this domain [19].

Neural networks and discriminant analysis are other techniques that allow us to predict the viability of banks. Neural networks show a better efficiency in their predictive ability than the former, especially because they weigh Type I errors more stringently. Predictive models successfully complement banks' on-site examinations as they are less intrusive and can be updated from time to time. But they do not provide a direct means for evaluating management and individual loans and underwriting practices [20].

4.2.1 Predictive Analysis of Stock Prices Using DCC GARCH Model in R

In one of our previous research works on the predictive analysis, we had conducted the DCC GARCH modeling of three companies in the stock market, namely, Netflix (NFLX), Disney (DIS), and Google (GOOGL). The statistical software is used for this analysis in R Studio. Figure 4.1 shows the GARCH multifit function generated in R.

Figure 4.2 indicates the information criteria such as Akaike, Bayes etc. Akaike criterion measures the likelihood of the estimated statistical models. It measures the quality of the model in relation to other models. Bayes information criterion is also used for evaluating a set of models. The low values of these information criteria for this model guarantee the feasibility of this model and indicate that it is useful for prediction [21].

Figure 4.3 shows the DCC modeling for these three companies for a period of previous 10 years starting from 2010 to 2019, which was conducted to analyze their correlations during this time.

The DCC between the three companies showed a positive correlation between NFLX and GOOGL as well as between GOOGL and DIS. This can be attributed to the fact that they are companies of different industries. In contrast, NFLX and DIS, which are competitors from the same industry, show a slightly negative correlation during 2018.

With the help of this historical data, the correlation forecast model was generated as shown in Figure 4.4. In the context of the predictive analysis, the correlation forecasts for a period of 10 upcoming years were done. This figure shows the correlation forecasts for the first two and the last two upcoming years. The same companies

```
> uspec.n<- multispec(replicate(3,ugarchspec(mean.model = list(armaOrder= c(1,0)))))
> multf= multifit(uspec.n, rX)
> multf

*----------------------------*
*      GARCH Multi-Fit       *
*----------------------------*
No. Assets :3
GARCH Multi-Spec Type : Equal
GARCH Model Spec
--------------------------
Model : sGARCH
Exogenous Regressors in variance equation: none

Mean Equation :
Include Mean :  1
AR(FI)MA Model : (1,d,0)
GARCH-in-Mean : FALSE
Exogenous Regressors in mean equation: none
Conditional Distribution:  norm

GARCH Model Fit
--------------------------
Optimal Parameters:
              rNFLX       rGOOGL         rDIS
mu          0.00210      0.00071      0.00062
ar1         0.04352      0.01183     -0.02745
omega       0.00000      0.00003      0.00002
alpha1      0.00788      0.08780      0.10523
beta1       0.99031      0.77735      0.81020
Log-Lik 5130.91387  6976.94039   7425.93495
```

FIGURE 4.1 GARCH multifit function.

```
*---------------------------------*
*          DCC GARCH Fit          *
*---------------------------------*

Distribution          :  mvnorm
Model                 :  DCC(1,1)
No. Parameters        :  20
[VAR GARCH DCC UncQ]  :  [0+15+2+3]
No. Series            :  3
No. Obs.              :  2515
Log-Likelihood        :  19912.28
Av.Log-Likelihood     :  7.92

Optimal Parameters
-----------------------------------
                    Estimate  Std. Error    t value
[rNFLX].mu          0.002105    0.000626    3.35956
[rNFLX].ar1         0.043524    0.025576    1.70172
[rNFLX].omega       0.000002    0.000001    2.26327
[rNFLX].alpha1      0.007876    0.000837    9.41195
[rNFLX].beta1       0.990312    0.000122 8106.17963
[rGOOGL].mu         0.000715    0.000309    2.31082
[rGOOGL].ar1        0.011828    0.022993    0.51444
[rGOOGL].omega      0.000034    0.000027    1.25624
[rGOOGL].alpha1     0.087797    0.071559    1.22692
[rGOOGL].beta1      0.777349    0.153274    5.07163
[rDIS].mu           0.000616    0.000306    2.01423
[rDIS].ar1         -0.027447    0.024512   -1.11973
[rDIS].omega        0.000015    0.000003    4.63766
[rDIS].alpha1       0.105235    0.015309    6.87408
[rDIS].beta1        0.810198    0.032688   24.78547
[Joint]dcca1        0.041001    0.016444    2.49343
[Joint]dccb1        0.860692    0.042913   20.05691
                   Pr(>|t|)
[rNFLX].mu         0.000781
[rNFLX].ar1        0.088808
[rNFLX].omega      0.023619
[rNFLX].alpha1     0.000000
[rNFLX].beta1      0.000000
[rGOOGL].mu        0.020843
[rGOOGL].ar1       0.606943
[rGOOGL].omega     0.209027
[rGOOGL].alpha1    0.219853
[rGOOGL].beta1     0.000000
[rDIS].mu          0.043985
[rDIS].ar1         0.262828
[rDIS].omega       0.000004
[rDIS].alpha1      0.000000
[rDIS].beta1       0.000000
[Joint]dcca1       0.012652
[Joint]dccb1       0.000000

Information Criteria
---------------------

Akaike       -15.819
Bayes        -15.773
Shibata      -15.819
Hannan-Quinn -15.802
```

FIGURE 4.2 DCC GARCH-fit.

depict the correlation of 1, which indicates a perfect correlation. The model shows a positive correlation (but it is less than 0.5)—it is a very low-degree correlation. So in upcoming years, the selected companies are predicted to showcase a slightly positive correlation. For reducing risk, it is suggested to diversify your investment portfolio, and this is done by choosing stocks which have low or negative correlations, so that in case one gives bad returns, you can depend on the other stock. In this way, banks can use the predictive analysis to gain a trading insight to guide their clients to buy,

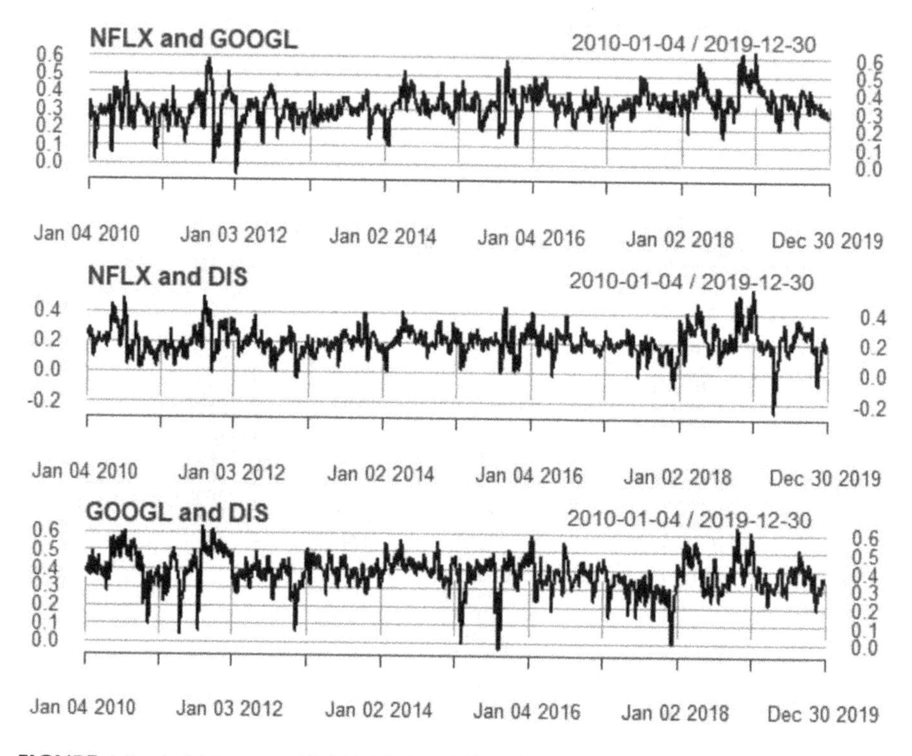

FIGURE 4.3 DCC between NFLX, DIS, and GOOGL.

```
*---------------------------------*
*         DCC GARCH Forecast      *
*---------------------------------*

Distribution     :  mvnorm
Model            :  DCC(1,1)
Horizon          :  10
Roll Steps       :  0
-----------------------------------

0-roll forecast:

First 2 Correlation Forecasts
$`1`

        [,1]   [,2]   [,3]
[1,] 1.0000 0.3174 0.2388
[2,] 0.3174 1.0000 0.3898
[3,] 0.2388 0.3898 1.0000

$`2`

        [,1]   [,2]   [,3]
[1,] 1.0000 0.3178 0.2352
[2,] 0.3178 1.0000 0.3885
[3,] 0.2352 0.3885 1.0000

. . .
```

```
. . .

Last 2 Correlation Forecasts
$`1`

        [,1]   [,2]   [,3]
[1,] 1.0000 0.3197 0.2180
[2,] 0.3197 1.0000 0.3826
[3,] 0.2180 0.3826 1.0000

$`2`

        [,1]   [,2]   [,3]
[1,] 1.0000 0.3199 0.2164
[2,] 0.3199 1.0000 0.3821
[3,] 0.2164 0.3821 1.0000
```

FIGURE 4.4 DCC-GARCH forecast model.

sell, or trade stocks. Predictive analysis coupled with the anomaly detection, which is explained later in this chapter, can also be used to detect future frauds or credit defaults as well.

4.3 GENETIC ALGORITHMS

Machine-based learning techniques have turned out to be very popular due to the face-paced development of information technology. Techniques such as neural networks and GAs are increasingly applied in the finance and investment field. Humans have cognitive abilities that are very limited in comparison with computers. With the help of technology and computers, more efficient trading strategies, financial scenarios, and portfolios can be developed. GAs are one such method for an efficient and quick evaluation of the real-time financial and investment possibilities. GAs are an effective technique for optimization. This technique is known as a stochastic optimization technique, which was developed by John Holland [22].

GA is an iterative for finding an optimum size, and it manipulates a population that is of a constant size. A population consists of chromosomes, and chromosomes consist of genes. These chromosomes compete with each other to survive, and in an applied scenario, it represents the potential solutions to the problem. The most optimum of these solutions will be left behind as survivors. Each iteration gives rise to a new population with the same number of chromosomes, which are better adapted to the selective function. Each new generation moves more closer to the optimum solution. Selection, crossover, and mutation are the operators that are applied to create the new population [23].

The operations of selection, crossover, and mutation help to converge or reduce the population into one optimal chromosome. They have been used in several financial applications, and they are found to be one of the most effective and most used approaches in finance such as for generating trading rules and portfolio formulation, optimizing crossover strategies etc. In the portfolio optimization, GA is employed to optimize the weight of each security or to decide the optimal proportion of funds that should be invested in each security. Each chromosome that indicates this weight is optimized.

GA follows the five steps: initialization, evaluation, selection, crossover, and mutation. GA generates the first set of solution, which is the initial population of chromosomes. The genes in the chromosome indicate the identification of stock. The value of these genes indicates the weight of each stock. The evaluation refers to the process of defining the fitness functions for the portfolio. To create a new chromosome, parent chromosomes are chosen by the proportion to the fitness value, and this is the process of selection. Crossover is a unique characteristic of GA in which the parent chromosomes are crossed over to obtain the offspring. It can be uniform or traditional. Uniform crossover involves a random selection of the gene, and it preserves schema as compared to traditional. Mutation alters the genes in the chromosome to ensure that the solution is not caught in a local optimum. The process is repeated multiple times to get the optimum solution or optimum portfolio. The iterative process is repeated until no further optimization or improvement is possible [24]. The use of GAs in the investment portfolio optimization is discussed in detail below.

4.3.1 Genetic Algorithms in Portfolio Optimization

Depending on the investor's tolerance to risk, active or passive portfolio strategies can be employed. Passive strategy is based on the Random Walk Theory, and it involves tracking the market index and is more of a passive or buy-and-hold strategy. Active strategy is employed by risk takers mostly, and it aims at outperforming a standard market index [24].

Portfolio optimization essentially means maximizing returns and minimizing risk of a portfolio. The risk level of a portfolio is often chosen based on the investor's tolerance of risk. This is the crux of Harry Markowitz's Modern Portfolio Theory. The mean–variance theory is the proposition that either an investor should minimize the risk for a particular level of return, or the vice versa. It makes a way for the efficient frontier, which is a graph showing all the feasible portfolios that are attainable, with different levels of risk and return.

The expected return of assets is the weighted average of the historical returns:

$$E[r_P] = \sum_{i=1}^{n} \omega_i E[r_i] \tag{4.1}$$

where $E[r_i]$ refers to the expected returns of individual stocks and ω_i is the weight of each stock in the portfolio, or the proportion of funds that are invested to each stock.

Risk of an asset or security is the uncertainty of its value in the future. It is measured using variance, standard deviation, semi-variance, VaR, CVaR etc. [25].

The risk is measured as follows:

$$\sigma_i^2 = \mathrm{Var}(r_i) = E\left[\left(r_i - E[r_i]\right)^2\right] \tag{4.2}$$

Covariance is measured as:

$$\sigma_{ij} = \mathrm{Covar}(r_{ij}) = E\left[\left(r_i - E[r_i]\right)\left(r_j - E[r_j]\right)\right] \tag{4.3}$$

Also, the average correlation between the stocks in a portfolio is:

$$\rho_{ij} = \frac{\sigma_{ij}}{\sigma_i \sigma_j} \tag{4.4}$$

Most investors opt for the low- or medium-risk portfolios, and people are generally averse to risk. To lower the risk, portfolio should have negative or low correlations. This helps to diversify risk [24].

Sharpe ratio measures the risk premium for an investment and is calculated as:

$$S_p = \frac{R_p - R_f}{\sigma_p} \tag{4.5}$$

where R_f is the return of a standard security, R_p denotes the return of the portfolio, and σ_p is the risk of that portfolio. A high Sharpe ratio indicates that the portfolio gives higher return for a particular risk. The weights of stocks that allow for maximizing the Sharpe ratio are known as optimal weights, and it is figured out for each stock using GA. Figure 4.5 shows a flowchart on the portfolio optimization using GAs.

In portfolio optimization of the objective function, we consider the total expected return and risk, which have to be maximized and minimized, respectively. The fitness function is defined as:

$$\text{fitness} = \frac{f(\text{return})}{f(\text{risk})} = \frac{\sum_{i=0}^{n} r_i \cdot w_i}{\sum_{i=0}^{n}\sum_{f=1}^{n} w_i \cdot w_j \cdot \sigma_{ij}} \quad (4.6)$$

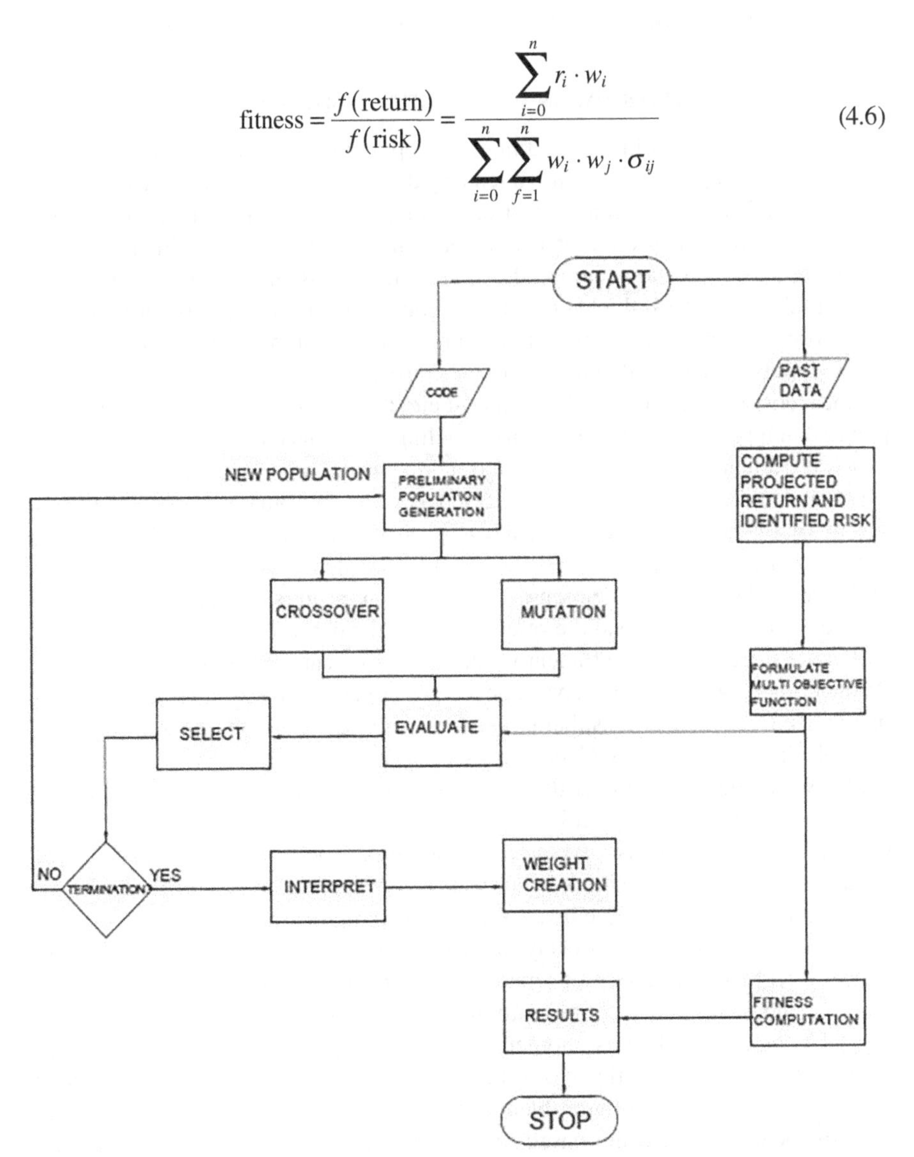

FIGURE 4.5 Process flow in a GA-based portfolio optimization.

GA ends by testing for convergence; that is, it gives results that converge to an optimal or better solution [26]. The advantage of using GA for multiobjective optimization is that it allows us to work with several portfolios and find multiple nondominated solutions in a single run itself. GAs are also less vulnerable comparatively to the nonconvex search space. But on the downside, they are time-consuming and the algorithm may have to be stopped before making any improvements to the efficient frontier, if the optimal solution is not known. A novel method in which the samples are split by quartiles of risk is discussed by J. Samuel et al. in 2012. This approach allows new space of solutions to be generated without increasing computation time [27].

4.3.2 Genetic Algorithms in Bank Profit Maximization

GAMCC (genetic algorithm based on multipopulation competitive coevolution) can be effectively employed to optimize lending decisions of the bank in order to maximize its profit in an environment of credit crisis. Traditional methods may not be effective for this because the objective and constraint functions in this problem are nonlinear and nonconvex, making the loan portfolio a complex problem to estimate. The variables on which the loan decision depends on are time period of loan, credit limit, size and type of the loan, its interest rate, and credit rating of the customer. The fitness function that should be maximized consists of the total transaction cost, loan revenue, cost of demand deposit, loan cost etc. In this case, GA generates the initial population by generating a collection of lending decisions on a random basis, each of which is evaluated with the assistance of a fitness function. The value of this fitness function decides whether it can be selected as a chromosome in the new population for the next generation. The next generation starts with reproduction: the best representation of the chromosome gets more copies, while the worst representations are excluded in the future generations. The algorithm ensures that the total amount of loan lent does not go beyond the reserve ratio of the deposit [28].

If banks are unable to efficiently manage their loan portfolios, it can lead to a credit crunch, which is the consequence of careless lending, causing losses and bad debts. In such situations, banks resort to increasing the interest rates for credit control. Banks can employ GAs to make an optimum decision when they have to cut back on lending supply; at the time, they face a negative liquidity shock, while focusing on the aim of maximizing their profit and minimizing the crediting cost. GA models can also be used to optimize the investment portfolio by making use of behavioral patterns like investor sentiments and overconfidence. This model helps to rationalize the decision of the investor in the stock market [29].

ATM deployers struggle to stay in business because demand in this field is less than the ATM supply and profit margins are under pressure. In order to meet the customer demand, banks should maximize the utility through effectively deploying ATM machines. This is another maximization problem in which GAs can be applied. Rank-based genetic algorithm using convolution (RGAC) can be employed to solve this ATM's location problem of banks, which is found to be more efficient than the heuristic algorithm based on convolution (HAC) because of the fact that the former allows for faster convergence while obtaining a feasible result. RGAC involves minimizing the difference between client utility (CU) and service utility

(SU), and computing the PC (percentage coverage) for each deployment to obtain the maximum PC. In this particular problem, gene indicates an ATM location, which may be 1 or 0. The fitness equation is devised to obtain the highest PC of CU. In this way, the RGAC allows banks to attain the highest CU and also improves their cost-efficiency [30].

Multiobjective selection genetic algorithm (MSGA) is also employed to identify the variables, which are most useful to banks and other financial institutions in predicting the user trust in e-banking systems. Non-dominated sorting genetic algorithm (NSGA-II) is quite expensive in terms of computation time, and also, it produces variables in very few numbers, which makes it difficult to be applied in real life. MSGA offers more complete solutions with significant variables enabling better managerial decisions. Hence, the variables selected by MSGA are more reliable in improving e-banking relative to other methods [31].

GAs are used in automatically assigning requests of customers to the respective helpdesk analysts. GAs identify which analyst is better suited to handle a particular customer request with the help of certain keywords associated with the analyst (which is updated regularly). The algorithm is designed to find the best solution analyst to which the customer request is automatically assigned to [32].

Credit institutions also are in need of better techniques for credit risk assessment as the industry now demands faster and less-risky operations on the precise data along with the reduced profit margin. Feature selection is another important aspect with regard to this, which is performed by hybrid genetic algorithm with neural networks (HGA-NN). HGA-NN heuristic can quickly remove irrelevant features and is a robust method added to current data mining techniques [33].

4.4 ANOMALY DETECTION

Anomaly detection is the process of finding patterns within a set of data that do not follow a normal or expected behavior. These patterns are referred to as outliers, or anomalies, exceptions, aberrations, peculiarities etc. depending on the area of application. This process is extensively used in the detection of fraud in credit card transactions, health care, insurance, intrusion detection for cybersecurity, and safety systems, and also in military surveillance. This area of data science is significant because anomalies or outliers in data often lead to critical information that needs to be immediately investigated and acted upon. Anomaly detection is distinct from noise removal and noise recommendation, which involves dealing with noise, which is something that poses as an obstacle in the way of data analysis. Novelty detection is another related area of work in which analysts aim to detect novel patterns in data, which were previously undetected. This is different from anomalies, because unlike anomalies, novelties are usually incorporated into the model after their detection. Anomaly detection is not an easy problem to solve in its general form, and hence, solving a specific formulation of the problem is more appropriate. Figure 4.6 shows the key concepts involved in an anomaly detection technique.

Anomalies are categorized as point, contextual, and collective anomalies. Point anomaly is the simplest; it is when an individual or point data is anomalous from the rest of the data. For example, if we have a data set regarding an individual's credit

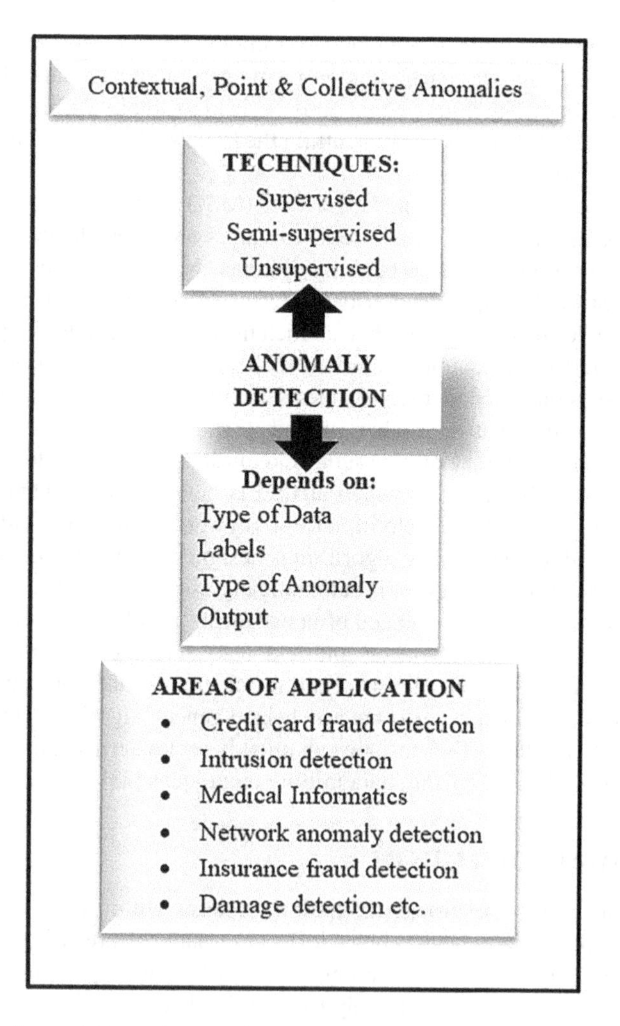

FIGURE 4.6 Anomaly detection—major concepts.

card transactions, we assume that the data is defined only by the amount spent. So a transaction which is carrying a rather high amount in comparison with the normal expenditure range will be a point anomaly and the particular transaction will have to be investigated further to check whether it is a fraud. Conditional or contextual anomaly is when a data instance is anomalous in a specific context alone; in a different context, the same data may not be anomalous. For example, in the case of credit card transactions, an individual has a usual shopping bill of US $ 100 but it increases to US $ 1000 during Christmas, which is considered normal; but a purchase of this same high amount in June will be a contextual anomaly. Collective anomaly refers to when a collection of data instances is found to be anomalous in comparison with the rest of the data. Their collective occurrence is the reason for them being considered as anomalous. The techniques used for detecting different types of anomalies are different; they can be supervised, semi-supervised, or unsupervised [34].

There are different approaches to anomaly detection. One is anomalous substructure detection, in which a graph is examined to identify anomalous data or outliers within it. This method is not very realistic in case there are very small substructures, and is effective for spotting large substructures. And large substructures are more infrequent than small substructures. The second method is anomalous subgraph detection in which the graph is divided into unique subgraphs, and it is evaluated to see how anomalous each subgraph is in comparison with others. Subgraphs that contain few common substructures are usually more anomalous than subgraphs that contain several common substructures. The concept of conditional entropy denotes the amount of information that is necessary to describe an event, if another particular event has occurred. It is a good measure for calculating the regularity or predictability of a graph. The regularity of the graph significantly influences the ability to detect anomalies within it; as the regularity increases, the average ranking of anomalies decreases [35].

An event or object is said to be anomalous if it deviates significantly from what is defined as normal, or that which is specified by the normality model. If S is a supervised anomaly detection approach, then it depends on two factors M and D, which are the normality model and degree of deviation, respectively. Therefore, each system will have a detection module and a modeling module. Anomaly detection is particularly useful in the detection of intrusion, which refers to actions that may breach the security of computers and networks. Anomalies in networks can be attributed to performance or security. Anomaly-based network intrusion detection (ANID) is a major area under research, and more systems with ANID capabilities are available now, but many essential areas need more clarity before such systems can be employed on a large scale [36].

Network anomalies can be attributable to issues like defect in device, network overload, malevolent denial-of-service attacks etc. One important requirement of detecting network anomalies is getting the right kind of network performance data. A greater synergy between signal processing areas and networking will allow the development of even more advanced tools in network anomaly detections. Signal processing tools make the existing network management tools more effective [37].

Unsupervised approach towards anomaly detection involves large data sets in which most elements are normal but there exist intrusions buried within the data. They assume that the normal events are very large when compared to the anomalies which are also qualitatively different. They are limited in the sense that they will not be able to identify those anomalies which are not qualitatively different from the normal events or instances. A geometric framework for unsupervised anomaly detection was suggested by Eskin et al. in 2002, in which data elements are mapped into a feature space and anomalies are identified by evaluating which elements occur in sparse regions of feature space. Feature space in the case of unsupervised anomaly detection is specific to its application; hence, it is best to choose the feature space after analyzing the application for optimal results. For detecting the points that lie in sparse regions, cluster-based algorithms, SVM approach, or k-nearest neighbor-based approach can be employed [38].

ANID is significant because web-based attacks are becoming more frequent and they should be addressed by techniques that have both precision and flexibility. Anomaly detection can also be performed by using input HTTP queries, to develop a system that is specifically made to detect web-based attacks. This model makes use of the correlation between server-side programs and their parameters which are specific to the application area. Extending this approach can enable to allow for the real-time anomaly detection for websites that address huge amounts of queries every day [39].

For different types of network attacks, different detection techniques can be employed and their success largely depends on the area of application. Unsupervised SVMs are very effective as they have a very high detection rate, but that said, they also have high false alarm rates. Development of anomaly detection techniques as such is an in-demand field because serial data mining algorithms prove to be less effective and unacceptable in the case of large volumes of heterogeneous data [40].

Traditional anomaly detection methods include GARCH model-based approach and OC-SVM model. GARCH, ARIMA, ARMA (autoregressive moving average), and EWMA (exponentially weighted moving average) are some of the early parametric anomaly detection methods. In this technique, we either assume a distribution for anomalies or follow a regression-based model. In the latter, the residual, which is the part that is not addressed by the regression model, is used to determine the scores for anomaly. SVMs, on the other hand, are nonparametric models, and they use local kernel models as opposed to a single global distribution model to the data. These models are faster, less complex, and at the same time flexible. This SVM model is a supervised approach, and it makes use of kernel function, so it has to be judiciously tuned for getting accuracy in classification [41].

Deep learning is used to analyze data sets in banking and insurance sectors to detect fraud. Algorithms such as AE (autoencoder), SVM, and deep convolutional networks are based on deep learning. Though several algorithms based on deep learning are available, developer implementing the technology should be aware of the problem in detail and also be aware on how each algorithm functions [42]. SVM models are associated with neural networks. Their aim is to find the optimal hyper-lane to separate the samples. The samples that are grouped near the hyper-lane are support vectors. OC-SVMs are different from these models because training data belongs only to one class [43].

Misuse of credit cards is a significant area in which anomaly detection can be highly useful. Technological developments have made money handling easier but at the same time, also made it easier to make fraudulent transactions and to produce forged cards. This may be done by getting hold of card details through a hidden device in ATMs or though shoulder surfing etc. Anomalies in credit card transaction can be identified by picking out transactions that involve very high payments or purchase of unusual items and also the increased frequency of purchase. Anomaly detection is also applied in identifying mobile phone frauds, by monitoring usage patterns and by creating a profile for each customer account, so that any deviation from the normal usage pattern of the customer can be detected as an anomaly and a subsequent warning can be issued to the customer to prevent any breach. Insurance frauds such as automobile-related frauds usually happen by forging documents. These are usually checked manually by insurance investigators, but detecting these frauds in an automated manner will be faster and even more efficient. Early detection is also crucial in the case of insider trading, where profit is made illegally out of insider information before it goes to the public. It is a criminal activity, which can affect the prices of stocks artificially.

A major challenge in the anomaly detection is the fact that it is not generalized; that is, the technique followed in one area or domain may not be suitable for another domain, even if the type of data is similar. Also, data contains noise, which may be perceived as an anomaly, making it difficult to distinguish. In addition, what is considered as normal behavior now may change in the future, so the current techniques of the anomaly detection may not be valid or useful in the future. As these techniques develop, fraudsters also become smarter, effectively imitating the normal behavior to make fraudulent activities normal [44].

Anomaly detection is especially important since invaders are becoming extremely smart in their attacks and financial sector around the world is facing huge data breaches using malicious tools of cyberattacks. It is becoming extremely necessary that Big Data Analytics is used to prevent and mitigate such challenges. This will allow financial institutions to live up to their commitment in protecting the sensitive data of their customers and stakeholders [45].

4.4.1 Anomaly Detection to Identify Credit Card Frauds using Python

4.4.1.1 Python Libraries

NumPy is an open-source module of python commonly used for computations related to arrays and matrices. High-level computations related to mathematical functions are performed using numpy. Pandas is similar to numpy, is used in python, and supports data analytical tools. Pandas provides a spreadsheet like rows and columns; hence, it is called as a dataframe. Plotting graphs and creating pivot tables were possible with the help of Pandas. Matplotlib in python is a two-dimensional plotting library and hence is used in Python scripts and web application servers [46].

Python uses data visualization library known as Seaborn that helps in providing an interface for drawing self-explanatory and useful statistical illustrations [47]. Scikit-learn is a python library that provides functionalities like regression, classification, clustering, model selection, and preprocessing, hence supporting supervised and unsupervised learning [48].

4.4.1.2 Anomaly Detection in Credit Card Data set

We have conducted the anomaly detection on a data set of credit card transactions, which was obtained from Kaggle.com. The data was subject to a univariate anomaly detection in Python, to identify the patterns that do not correspond to the normal behavior.

The distribution of variable "Amount" which we have obtained is very different from the normal distribution as the most of the distribution concentrates itself on the left side of the graph. The data has very low probability to appear on the right side of the distribution. The distribution is highly skewed; it is calculated to be a positive value of 16.49, indicating that the right tail of the distribution is longer than the left as shown in Figure 4.7. The distribution is leptokurtic; kurtosis is calculated as 672.963, indicating extreme values in the distribution.

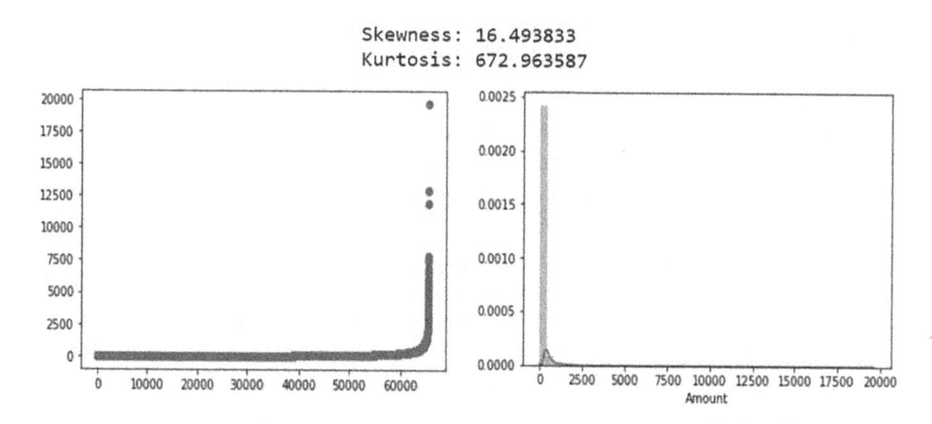

FIGURE 4.7 Distribution of amount of transaction.

The algorithm that we have used here to detect the anomalies is Isolation Forest, a tree-based model. Figure 4.8 shows the region where the outliers fall. According to the graph, transactions that exceed 500 are considered as an outlier, and further investigation needs to be put into such transactions to evaluate whether they are fraud or not. With the help of this model, we are also able to visually investigate each anomaly.

4.4.2 A Demonstration of Anomaly Detection in Ethereum Prices Using R

4.4.2.1 Ethereum

Ethereum can be considered as a transaction-based state machine, which contains data such as reputations, account balances, data relating to physical world etc. Transactions are collated into blocks that are joined together by means of cryptography. The blocks work as a journal or ledger, chaining together these transactions

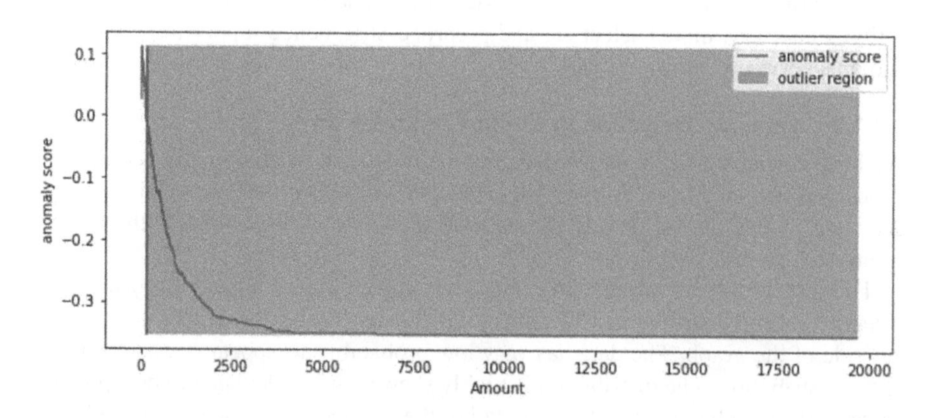

FIGURE 4.8 Univariate anomaly detection of transaction amounts.

with the preceding block. The decentralized system allows all parties involved to create new blocks [49]. Bitcoin and Ethereum are the most popular cryptocurrencies available today. Ethereum aims to overcome several of Bitcoin's limitations. It supports all kinds of computations and loops included; it also supports the state of the transaction and has many additional improvements compared to Bitcoin's structure. It allows anyone to create their own rules for ownership by using smart contracts, which can be executed only by meeting certain conditions [50].

Ethereum and other cryptocurrencies stay as a fast, secure layer on top of the existing currency system, and it is an invention that has the capability of replacing the centralized payment systems in place today. Ethereum is suitable for constructing economic systems purely in software. It enables people to move money with the ease and speed with which you move data today. It eliminates floating periods and fees associated with vendors such as Mastercard, Visa etc. Ethereum preserves several of the fundamental concepts of Bitcoin, but it is altogether new, as its key components differ. It can be replicated into compatible systems; it is a free and open-source system [51].

4.4.2.2 Tidyverse

Tidyverse is a collection of R packages that are featured by low-level grammar and data configurations and high-level design. It is an easier way of facilitating repeated tasks such as importing large amounts of data, tidying it, manipulating, visualizing, and programming of this data. It allows us to download and install all tidyverse packages using a single command; this single metapackage that enables this is called tidyverse. Tidyverse works with tidy data, in which each row is an observation and each column is a variable. Each cell has a single value. Tidyr allows you to convert your data into this tidy format, making it easier to work with. Transforming this data is done with the help of dplyr package. Visualization of this data is done with the help of graphs, and modeling is done using tidymodels. Communicating this data with others is not included in tidyverse package, and it is done with the help of other packages in R such as rmarkdown and shiny. Programming surrounds all these above-mentioned aspects and is cross-cutting to every aspect of data science.

Purr, tibble etc. are programming tools in tidyverse package [52]. The elements of Tidyverse package are shown in Figure 4.9.

4.4.2.3 Anomaly Detection

We collected historical data of Ethereum prices from 2018 from Coindesk.com and conducted the anomaly detection using tidyverse in R. First, we decompose the time-series data of Ethereum prices into seasonal, trend, and remainder components. Then, we use "Anomalize" library in R to detect and flag the anomalies. The white points in the graph indicate anomalies that need to be investigated further, while the black points in the graph indicate the normal behavior. We can also use R to extract the actual points, which are anomalies.

The time-series decomposition and anomalies detected are shown in Figures 4.10 and 4.11.

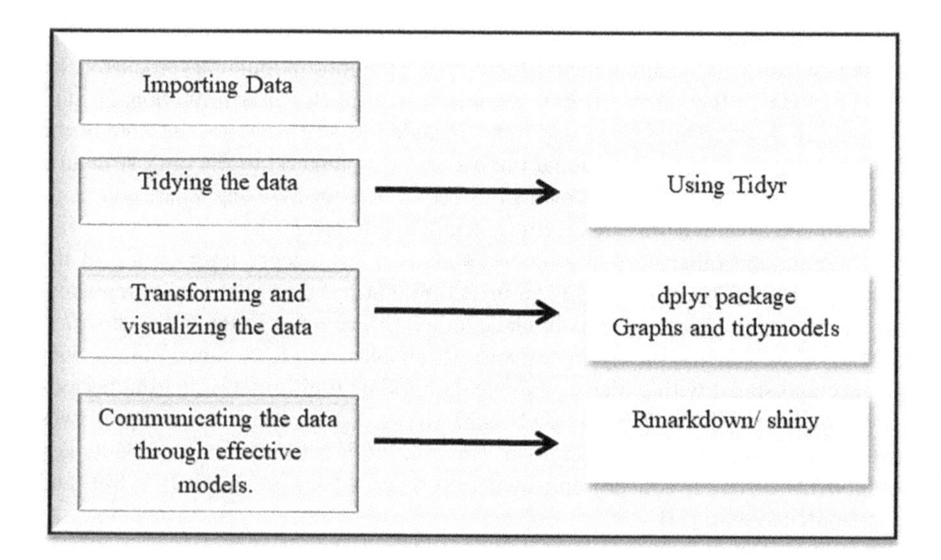

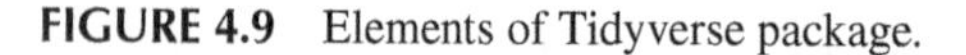
FIGURE 4.9 Elements of Tidyverse package.

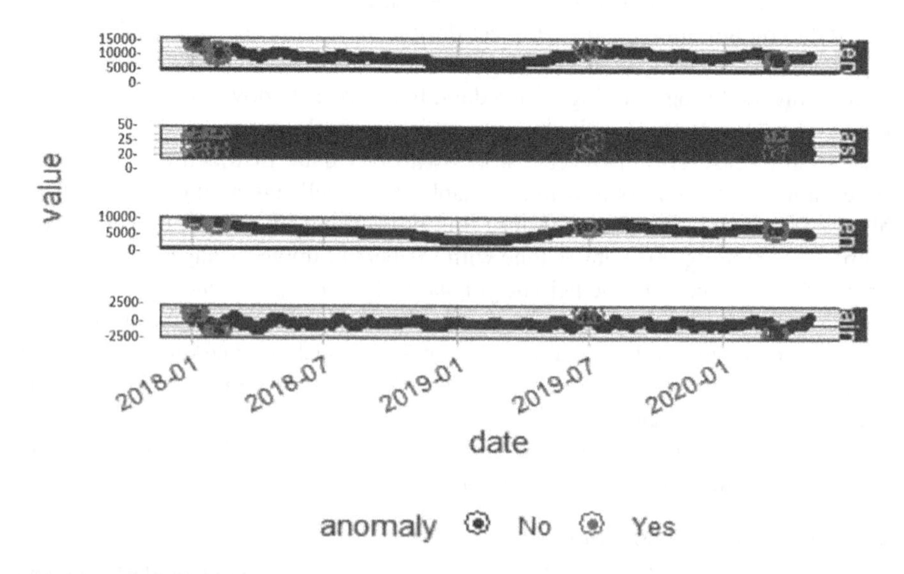

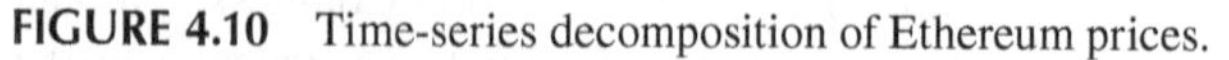
FIGURE 4.10 Time-series decomposition of Ethereum prices.

4.5 CONCLUSION

AI is the biggest technological revolution now, and it is influencing and is expected to continue to influence the way in which companies carry out their processes. AI makes it possible to make use of enormous amounts of data that is generated day-by-day to benefit corporates to act in a more efficient and effective manner. Artificial intelligence has significant applications in the domain of insurance and banking, some of which we have discussed in this chapter include predictive analysis, GAs,

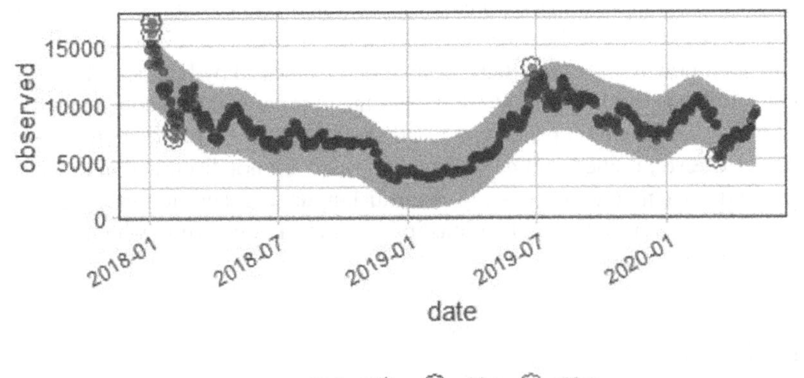

FIGURE 4.11 Plot of detected anomalies in Ethereum prices.

and anomaly detection. There are many other useful and robust applications as well. These techniques allow banks and other financial institutions to optimize their procedures and respond to customer demands more efficiently.

Predictive analysis allows organizations to obtain information about trends and patterns in historical data to understand the projected future. It is used by organizations to identify the trends in customer behavior and buying patterns, forecasting sales, quality management etc. In the domain of insurance and banking, predictive analysis is especially used to predict the customer churn and analytics, credit risk assessment, to obtain the trading insight etc. The application of predictive analysis in the trading insight is discussed in this chapter; the predictive analysis allows banks to guide their clients as to whether to sell, hold, or buy stock. A demonstration of DDC GARCH forecast model for a period of upcoming 10 years has been made between NFLX, DIS, and GOOGL using statistical software R. Results show a slightly positive correlation in the upcoming years for the selected companies. Correlation is a significant measure that allows investors to diversify their investment portfolio. Banks also use the predictive analytics to predict credit risk and loan repayment defaults with the help of SVMs, logistic regression models etc., which are made easier with the help of AI and computational software packages.

GAs help generate high-quality solutions for complex optimization problems. The repetitive process allows banks to choose the most optimal way of carrying out a process or an operation. GAs allow banks to maximize the profit by optimizing different areas such as deployment of ATM machines, loan portfolio optimization, investment portfolio optimization etc. They also enable better customer service and optimize the way in which analysts respond to customer queries.

Anomaly detection discussed towards the end of this chapter is an important aspect pertaining to credit card transactions and mobile banking; it enables banks to detect potential frauds. It involves identifying outliers or anomalies in large sets of data, which may be potential frauds. Identifying and preventing frauds are important for any organization, while for banks and other financial institutions, it is crucial, as they deal with large amounts of money and customers need to trust the institution in which they deposit their money with. Potential frauds or breaches can destroy customer's trust and adversely affect the long-term reputation of a financial institution. We have demonstrated how the anomaly detection can be employed by banks to detect frauds in credit card transactions using Python. Another demonstration is conducted to show the anomalies in Ethereum prices from 2018, using Tidyverse package in R.

There are several other applications of AI, which are not discussed in this chapter, such as natural language processing, natural language generation etc. These techniques also have significant applications in insurance and banking, which are beyond the scope of this chapter.

REFERENCES

1. Hall, S., 2017. How artificial intelligence is changing the insurance industry. *The Center for Insurance Policy & Research*, 22, pp. 1–8.
2. Alzaidi, A.A., 2018. Impact of artificial intelligence on performance of banking industry in Middle East OCR. *International Journal of Computer Science and Network Security*, 18(10), pp. 140–148.
3. Jewandah, S., 2018. How artificial intelligence is changing the banking sector – A case study of top four commercial Indian banks. *International Journal of Management, Technology and Engineering*, 8(7), pp. 525–530.
4. Pau, L.F. and Gianotti, C., 1990. Applications of artificial intelligence in banking, financial services and economics. In Pau, L.F. and Gianotti, C. (eds.) *Economic and Financial Knowledge-Based Processing* (pp. 22–46). Springer, Berlin, Heidelberg.
5. Vedapradha, R. and Ravi, H., 2018. Application of artificial intelligence in investment banks. *Review of Economic and Business Studies*, 11(2), pp. 131–136.
6. Zarifis, A., Holland, C.P. and Milne, A., 2019. Evaluating the impact of AI on insurance: The four emerging AI-and data-driven business models. *Emerald Open Research*, 1(15), p. 15.
7. McCue, C., 2014. *Data Mining and Predictive Analysis: Intelligence Gathering and Crime Analysis*. Butterworth-Heinemann, Oxford.
8. Waller, M.A. and Fawcett, S.E., 2013. Data science, predictive analytics, and big data: A revolution that will transform supply chain design and management. *Journal of Business Logistics*, 34(2), pp. 77–84.
9. Shmueli, G. and Koppius, O.R., 2011. Predictive analytics in information systems research. *MIS Quarterly*, 35(3), pp. 553–572.
10. Andersen, T.G., Bollerslev, T., Christoffersen, P.F. and Diebold, F.X., 2006. Volatility and correlation forecasting. *Handbook of Economic Forecasting*, 1, pp.777–878.
11. Erb, C.B., Harvey, C.R. and Viskanta, T.E., 1994. Forecasting international equity correlations. *Financial Analysts Journal*, 50(6), pp. 32–45.
12. Boudt, K., Danielsson, J. and Laurent, S., 2013. Robust forecasting of dynamic conditional correlation GARCH models. *International Journal of Forecasting*, 29(2), pp. 244–257.
13. Engle, R., 2002. Dynamic conditional correlation: A simple class of multivariate generalized autoregressive conditional heteroskedasticity models. *Journal of Business & Economic Statistics*, 20(3), pp. 339–350.
14. Chou, R.Y., Wu, C.C. and Liu, N., 2009. Forecasting time-varying covariance with a range-based dynamic conditional correlation model. *Review of Quantitative Finance and Accounting*, 33(4), p. 327.
15. Sen, J. and Chaudhuri, T.D., 2016. A framework for predictive analysis of stock market indices: A study of the Indian Auto Sector. arXiv preprint arXiv:1604.04044.

16. Oh, C. and Sheng, O., 2011. Investigating predictive power of stock micro blog sentiment in forecasting future stock price directional movement. In *Proceedings of the International Conference on Information Systems*, ICIS.
17. Smailović, J., Grčar, M., Lavrač, N. and Žnidaršič, M., 2013, July. Predictive sentiment analysis of tweets: A stock market application. In *International Workshop on Human-Computer Interaction and Knowledge Discovery in Complex, Unstructured, Big Data* (pp. 77–88). Springer, Berlin, Heidelberg.
18. Avramov, D., 2002. Stock return predictability and model uncertainty. *Journal of Financial Economics*, 64(3), pp. 423–458.
19. Kolari, J., Glennon, D., Shin, H. and Caputo, M., 2002. Predicting large US commercial bank failures. *Journal of Economics and Business*, 54(4), pp. 361–387.
20. Swicegood, P. and Clark, J.A., 2001. Off-site monitoring systems for predicting bank underperformance: A comparison of neural networks, discriminant analysis, and professional human judgment. *Intelligent Systems in Accounting, Finance & Management*, 10(3), pp. 169–186.
21. Javed, F. and Mantalos, P., 2013. GARCH-type models and performance of information criteria. *Communications in Statistics-Simulation and Computation*, 42(8), pp. 1917–1933.
22. Ranković, V., Drenovak, M., Stojanović, B., Kalinić, Z. and Arsovski, Z., 2014. The mean-value at risk static portfolio optimization using genetic algorithm. *Computer Science and Information Systems*, 11(1), pp. 89–109.
23. Guennoun, Z. and Hamza, F., 2012. Stocks portfolio optimization using classification and genetic algorithms. *Applied Mathematical Sciences*, 6(94), pp. 4673–4684.
24. Cheong, D., Kim, Y.M., Byun, H.W., Oh, K.J. and Kim, T.Y., 2017. Using genetic algorithm to support clustering-based portfolio optimization by investor information. *Applied Soft Computing*, 61, pp. 593–602.
25. Roudier, F., 2007. Portfolio optimization and genetic algorithms. Master's thesis, Department of Management, Technology and Economics, Swiss Federal Institute of Technology (ETM), Zurich.
26. Zuhal, L.R., 2010, February. Resolving multi objective stock portfolio optimization problem using genetic algorithm. In *2010 The 2nd International Conference on Computer and Automation Engineering (ICCAE)* (Vol. 2, pp. 40–44). IEEE, Singapore.
27. Baixauli-Soler, J.S., Alfaro-Cid, E. and Fernandez-Blanco, M.O., 2012. A naïve approach to speed up portfolio optimization problem using a multiobjective genetic algorithm. *Investigaciones Europeas de Dirección y Economía de la Empresa*, 18(2), pp. 126–131.
28. Metawa, N., Hassan, M.K. and Elhoseny, M., 2017. Genetic algorithm based model for optimizing bank lending decisions. *Expert Systems with Applications*, 80, pp. 75–82.
29. Metawa, N., Elhoseny, M., Hassan, M.K. and Hassanien, A.E., 2016, December. Loan portfolio optimization using genetic algorithm: A case of credit constraints. In *2016 12th International Computer Engineering Conference (ICENCO)* (pp. 59–64). IEEE, Cairo, Egypt.
30. Alhaffa, A., Al Jadaan, O., Abdulal, W. and Jabas, A., 2011, March. Rank based genetic algorithm for solving the banking ATM's location problem using convolution. In *2011 IEEE Symposium on Computers & Informatics* (pp. 6–11). IEEE, Kuala Lumpur, Malaysia.
31. Liébana-Cabanillas, F., Nogueras, R., Herrera, L.J. and Guillén, A., 2013. Analysing user trust in electronic banking using data mining methods. *Expert Systems with Applications*, 40(14), pp. 5439–5447.
32. Ciurea, C., 2011. Using genetic algorithms for building metrics of collaborative systems. *Informatica Economica*, 15(1), p. 80.
33. Oreski, S. and Oreski, G., 2014. Genetic algorithm-based heuristic for feature selection in credit risk assessment. *Expert Systems with Applications*, 41(4), pp. 2052–2064.

34. Chandola, V., Banerjee, A. and Kumar, V., 2009. Anomaly detection: A survey. *ACM Computing Surveys (CSUR)*, 41(3), pp. 1–58.
35. Noble, C.C. and Cook, D.J., 2003, August. Graph-based anomaly detection. In *Proceedings of the Ninth ACM SIGKDD International Conference on Knowledge Discovery and Data Mining* (pp. 631–636). ACM, New York, NY.
36. Bhuyan, M.H., Bhattacharyya, D.K. and Kalita, J.K., 2013. Network anomaly detection: Methods, systems and tools. *IEEE Communications Surveys & Tutorials*, 16(1), pp. 303–336.
37. Thottan, M. and Ji, C., 2003. Anomaly detection in IP networks. *IEEE Transactions on Signal Processing*, 51(8), pp. 2191–2204.
38. Eskin, E., 2000. Anomaly detection over noisy data using learned probability distributions. In *Proceedings of the Seventeenth International Conference on Machine Learning* (pp. 77–101). Morgan Kaufmann Publishers Inc. .
39. Kruegel, C. and Vigna, G., 2003, October. Anomaly detection of web-based attacks. In *Proceedings of the 10th ACM conference on Computer and Communications Security* (pp. 251–261). ACM Press, New York, NY.
40. Lazarevic, A., Ertoz, L., Kumar, V., Ozgur, A. and Srivastava, J., 2003, May. A comparative study of anomaly detection schemes in network intrusion detection. In *Proceedings of the 2003 SIAM International Conference on Data Mining* (pp. 25–36). Society for Industrial and Applied Mathematics.
41. Davis, N., Raina, G. and Jagannathan, K., 2019. A framework for end-to-end deep learning-based anomaly detection in transportation networks. arXiv preprint arXiv: 1911.08793.
42. Pumsirirat, A. and Yan, L., 2018. Credit card fraud detection using deep learning based on auto-encoder and restricted Boltzmann machine. *International Journal of Advanced Computer Science and Applications*, 9(1), pp. 18–25.
43. Sundarkumar, G.G. and Ravi, V., 2015. A novel hybrid undersampling method for mining unbalanced datasets in banking and insurance. *Engineering Applications of Artificial Intelligence*, 37, pp. 368–377.
44. Ahmed, M., Mahmood, A.N. and Islam, M.R., 2016. A survey of anomaly detection techniques in financial domain. *Future Generation Computer Systems*, 55, pp. 278–288.
45. Cyriac, N.T. and Sadath, L., 2019, November. Is cyber security enough – A study on big data security breaches in financial institutions. In *2019 4th International Conference on Information Systems and Computer Networks (ISCON)* (pp. 380–385). IEEE, Mathura, India.
46. Python libraries [Online]. https://cloudxlab.com/blog/numpy-pandas-introduction/ [Accessed on: 13-05-2020].
47. Seaborn [Online]. https://seaborn.pydata.org/ [Accessed on: 13-05-2020].
48. Scikit-learn [Online]. https://www.codecademy.com/articles/scikit-learn [Accessed on: 13-05-2020].
49. Wood, G., 2014. Ethereum: A secure decentralised generalised transaction ledger. *Ethereum Project Yellow Paper*, 151(2014), pp. 1–32.
50. Vujičić, D., Jagodić, D. and Ranđić, S., 2018, March. Blockchain technology, bitcoin, and Ethereum: A brief overview. In *2018 17th International Symposium INFOTEH-JAHORINA (INFOTEH)* (pp. 1–6). IEEE, East Sarajevo, Bosnia-Herzegovina.
51. Dannen, C., 2017. *Introducing Ethereum and Solidity* (Vol. 1). Apress, Berkeley, CA.
52. Wickham, H., Averick, M., Bryan, J., Chang, W., McGowan, L., François, R., Grolemund, G., Hayes, A., Henry, L., Hester, J. and Kuhn, M., 2019. Welcome to the Tidyverse. *Journal of Open Source Software*, 4(43), p. 1686.

5 Artificial Intelligence in Robotics and Automation

Udit Pratap Singh
Chandigarh University

Astha Mishra
Ajay Kumar Garg Engineering College

CONTENTS

5.1 INTRODUCTION

The term "artificial intelligence" can be observed as one of the most finest or one of the most unfavorable things for the mankind. AI is a focused area of science and technology to make machine knowledgeable, which basically means generalized learning, reasoning, analyzing, and understanding of natural languages. These days, the term "AI" envelops the entire concept of a machine that astute in wording of both operational and social results. When the AI is incorporated with robots, there comes a concept of automated bots, and we need to understand that it's not necessary that robotics always mean to have the physical robots; these can be application bots that work on human instructions and are made to mimic human work and automate the process and from here, the "automation" term is coined, which basically means the replicating human task and integrating these automated bots with AI to make decisions. On the one hand, automation leads to business benefits like cost, time, and a rapid production rate; on the other hand, the automated technologies lead to "technological unemployment." These technologies have not only suppressed the physical strength of human but also suppressed the human cognitive ability to work and process the huge amount of data and decision-making power.

The development of AI and automated technologies can lead a world to better off, but as these technologies are growing, they can lead to a negative impact on society if not handled accordingly. This chapter will discuss the features of automated bots, difference between bots and robots, features and scope of automation, and the technologies that support their implementation, and some of the vulnerabilities.

5.2 HISTORY

The term "automation" was coined in 1946 by D.S. Harder of Ford Motor industries, and it was used to describe the enhancement in the production line, which basically means replacing human workers with machinery. Later, in 1956, Dartmouth conference organized by Marvin Minsky, John McCarthy, and two senior scientists Claude Shannon and Nathan Rochester of IBM used the term "artificial intelligence." At that time, nobody has expected that it was possible to have an intelligent machine.

But as the time passed, the physical automated machinery came into existence, and as further development was done, software bots came into existence in 1988, which were only used to keep the server running even after inactivity. In 1994, the bot was created for indexing of web pages, which was basically called as WebCrawler. And later, Business Process Management (BMP) started growing, and by the end of 1990s, large BPM systems were implemented. These technologies and some other technology like natural language processing (NLP) are integrated together to make technology called Robotics Process Automation (RPA).

5.3 AUTOMATION AND APPLICATION BOTS

Automation is a technique in which process is carried out without human intervention or with a minimum human intervention; it can be in the form of physical bots or application bots. In this chapter, we will be focusing on the application bots only. So these bots are basically a set of instructions for carrying out a particular operation (generally BPM). These automated bots can become a complete machine once they are able to perform all the operations (i.e., the complete process). These automated bots select data, analyze it, and take decisions based on the analysis.

Automated bots were mainly designed to perform the repetitive rule-based task, and generally, the procedure of these tasks doesn't change over a short period of time. The main reason to design these bots was to engage the employees on a more productive and creative task rather than a reparative task. QR code scanner is one of the best examples of software bots though it doesn't look like a bot but it just scans a QR code and all the data will be recorded in the database and save the time of humans for saving all the data into the computer system manually.

Chatbots are one of the popular things which we all have heard about; they also look like other automated bots but there's a thin line of difference between robots, chatbots, and bots.

5.4 ROBOTS VS. CHATBOTS VS. BOTS

Robots: These are the devices that work automatically without human interventions. These generally resemble the physical structure of human being or some other creatures. But these are physically present in a dedicated manner to perform any particular task. These bots are popularly used in an industrial assembly line like automobile industry.

Chatbots: These bots are used to perform the human interaction, which can result to enhance the customer experience and can act like a help desk.

Bots: Bots are basically designed to work on a repetitive task basically on software, as per the definition but as the technology is growing, automated bots have been developed, which not only perform a repetitive task but also made capable to take decisions based on the processed data and perform further operations.

For the deeper understanding of AI with application bots, it is important to know how the communication is established between the bots and the humans, and for this purpose, NLP is used.

5.4.1 Types of Bots

Attended Bots: These bots are not completely automated; these bots require some manual interventions from humans.

Unattended Bots: These bots are completely automated, and they don't require human interventions.

For the deeper understanding of AI with application bots, it is important to know how the communication is established between the bots and the humans, and for this purpose, NLP is used.

5.5 NATURAL LANGUAGE PROCESSING (NLP)

NLP is used to perform an interaction between humans and machines. It converts the human language to machine-understandable form, and vice versa.

It basically works in two phases.

5.5.1 Natural Language Understanding (NLU)

This involves the process of understanding the input from the humans, by means of the following three ways:

- Lexical
- Syntactical
- Referential.

But there exists a problem of ambiguity when the words or sentences are interpreted by NLP.

Example

Lexical is used to process at the word level: for example, "alert" can be quick to notice (noun) as well as it can be a state of being watchful (adjective), so here can occur the problem of ambiguity.

Syntactical is used to work on a sentence level: for example, "Chicken is ready to eat." It can have two meanings: either chicken is cooked and is ready to eat or chicken is ready to be fed, so again it leads to ambiguity.

Referential sentences are the most complex one: for example, "Rita is sitting with Sita, and she is styling her hair," so here it can be confusing for the program to understand to whom "she" is referring to either Rita or Sita. This process also becomes an ambiguity problem.

So, depending upon the word and the sentence formation, the interpretation may differ, and this may lead to a wrong output in the next phase. This is the reason why sometimes bots don't give the relevant outputs to our inputs.

As the machines have understood the command in the next phase, they will process the output.

5.5.2 Natural Language Generation

Based on the input, the output is generated in a human-interpreted format. This process consists of many processes, which can be broadly classified as:

- **Text Planning**: Knowledge base is used to decide which words are needed to be selected for the output.
- **Sentence Planning**: Sentence is formed from the selected words in a meaningful way.
- **Text Realization**: In this process, the final output is given to the user [1].

The process of NLP is shown in Figure 5.1.

5.6 ROBOTICS PROCESS AUTOMATION (RPA)

RPA is used to automate business process by using various technologies of AI. RPA utilizes programming robots to do activities, for example, scraping, information conglomeration, information cleaning, and communication with different applications and individuals to execute a dreary work. It permits organizations to keep up low expenses while giving their employees an expanded chance to handle different needs inside their associations. RPA can likewise mechanize various dreary and tedious assignments. The tasks that are automated may or may not use AI based on the condition whether they are rule-based and repetitive task or decision-based task [2]. Figure 5.2 shows the schematics of RPA.

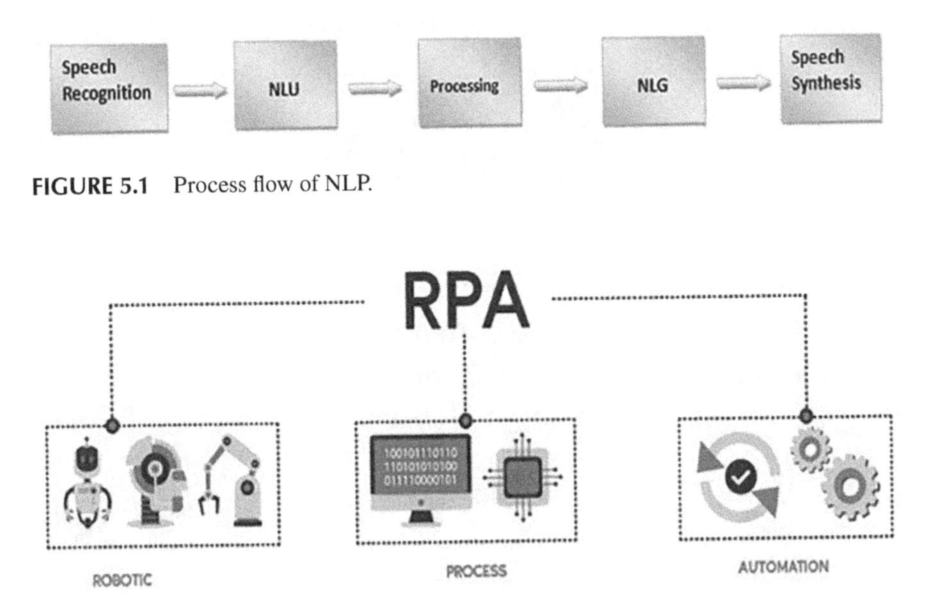

FIGURE 5.1 Process flow of NLP.

FIGURE 5.2 Robotic process automation.

RPA categorizes bots into three types:

1. **Task Bots**: These bots are the most basic bots in the field of automation; these bots generally perform the repetitive-based and rule-based tasks; they can easily perform a multistep operation but these bots can't work on semi-structured or unstructured data.
2. **Meta Bots**: These bots are used to automate tasks based on the desktop application; these bots are generally made by using visual captures.
3. **IQ Bots**: These are the intelligent bots and have cognitive capabilities. These bots can make decisions based on the analysis of data. These bots can work with semi-structured or unstructured data.

The various types of bots in RPA are shown in Figure 5.3.

RPA uses various AI technologies to automate business processes. Some of the most frequently used AI technologies are as follows:

- **Machine Learning (ML)**: It helps in data and pattern analysis, and makes decision with or without human intervention. With the help of ML, the system can learn from the data. As ML can't deal with high dimensional data, we need to use deep learning technology. The limitations of ML to solve the problems are further feature extraction or image recognition, or hand writing recognition. Hence, in this case, deep learning is implemented.
- **Deep Learning**: Deep learning is a technique based on AI that emulates the human brain working to process data and generate patterns to take a decision on the problem. Deep learning has the property of unsupervised

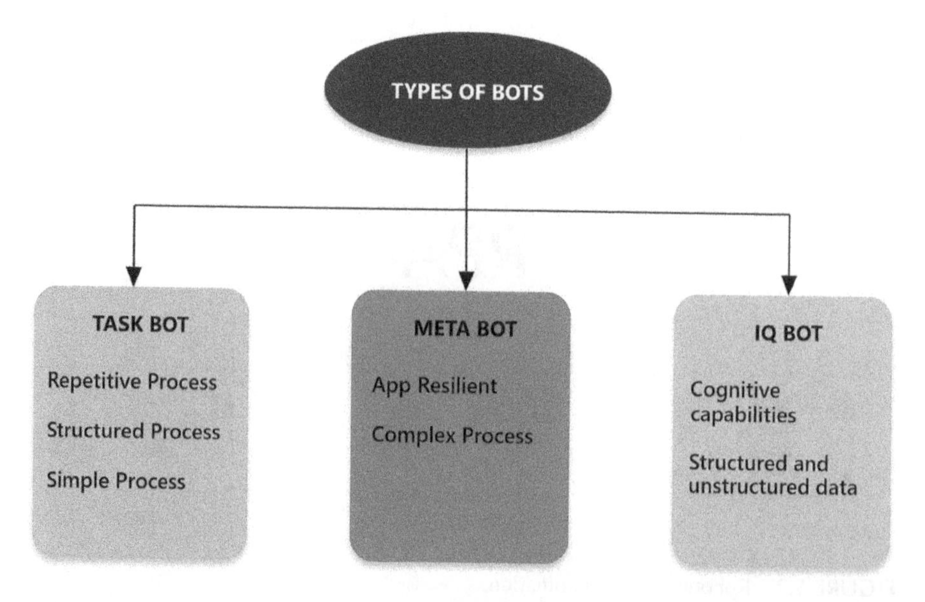

FIGURE 5.3 Types of bots in RPA.

learning from unstructured and unlabeled data. The data is basically drawn from search engines, social media platforms, and e-commerce platforms.
- **Image Recognition**: The technology recognizes and identifies the items or qualities in pictures or videos.
- **Speech Recognition**: The innovation recognizes words and expressions in communication language, and changes over them into a machine-meaningful arrangement.

Even though we have a huge list of technologies present with us, still we can't automate each and every task so we need to understand which tasks can be made automated with the bots, which tasks still require human intervention, or which processes can't be automated at any point. Great understanding of process is required to decide which part of process can be automated and up to what extent.

5.6.1 Challenges in Implementation of RPA

1. Process Selection
   - Difficult to automate higher-level process
   - Lack of process standardization
2. Availability of Relevant and Sufficient Talent
   - Shortage of skills
   - Lack of knowledge to implement AI
3. Organizational Readiness
   - Managing organizational changes
   - Concern from IT
4. Lack of Sufficient Data to Train AI
   - Extracting data and making it relevant to train AI
5. Implementation
   - Scaling up smart RPA
   - Business people are lacking smart RPA programming skills, and smart RPA developers are lacking business contexts [3].

Figure 5.4 shows various challenges in the implementation of RPA.

5.7 FINANCIAL IMPACT OF AI AND AUTOMATION

As business process automation is saving cost for the organization, as well decreasing the time for the execution of process, its demand in the industry is growing rapidly. The pandemic COVID-19 made many organizations to think about the bots rather having a full-time employee (FTE). Previously for many small-scale organizations and firms, bots seem to be costly, but due to downfall of economy, many organizations are struggling to pay their employees and they have started to shift towards the automation at least for those repetitive tasks which were done by an FTE. Automated bots seem to be costly, but when it comes to longer run and repetitive tasks, these bots are best to deploy as the organizations have to pay only for once and a single bot can replace the work of multiple employees.

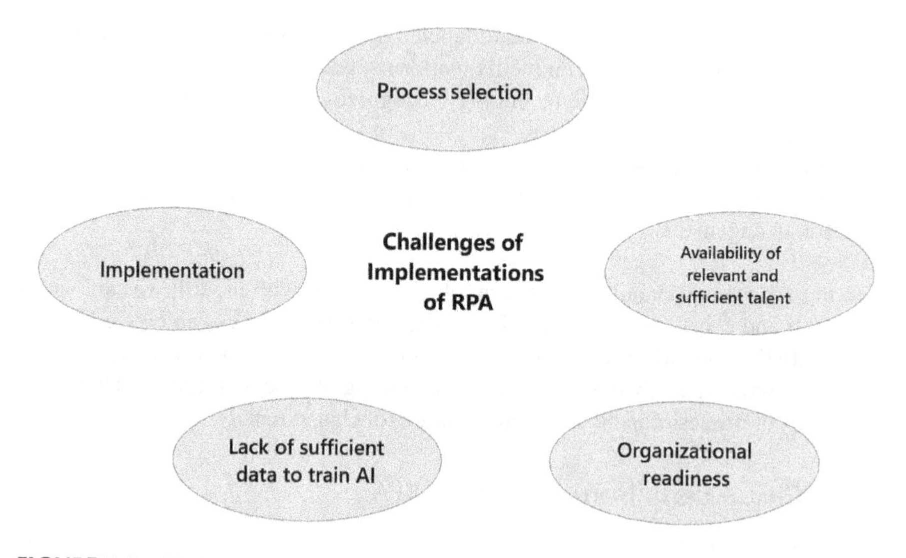

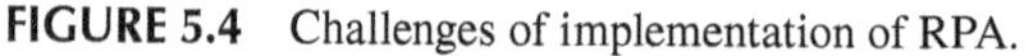
FIGURE 5.4 Challenges of implementation of RPA.

As per the market forecast, the revenue for RPA market is estimated around US $ 214 billion worldwide in 2021, and RPA is being considered as one of the most demanding jobs in IT industry. Every organization wants to maximize their profit and involve their employees in more productive jobs.

Organizations like Microsoft, Google, Facebook, Amazon, and IBM are investing a huge amount in the field of development of AI technologies and also supporting to develop more efficient and intelligent bots. Some of the major investments are done in the field of speech reorganization, NLP, computer vision, optical character recognition, deep learning, and Big Data. Bots are being made more efficient to work on semi-structured and unstructured data.

In 2018, Google acquired customer service-based startup "Onward"; the motive of the startup was to enhance the human–computer conversation.

IBM has developed a super computer for the purpose of data mining and analyzing the pattern from highly unstructured data. AlchemyAPI is helping to develop intelligent bots to process text as well as images. These systems are already helping to process legal documents and in the health care system.

In 2014, Google acquired "DeepMind," an AI-based startup, and the company showed a remarkable growth in the field of general-purpose software that adjusts their working based upon cumulative rewards. This startup builds "WaveNet" that can mimic human voice and enhance user experience.

Robo-advisor is one of the most demanding bots and is capable of performing NLP and interaction with humans; these bots create a revolution in the customer service market; we all are very familiar for using these technologies (e.g., Amazon Alexa, Google Assistant, Siri for apple). In 2016, Apple Inc. acquired a startup that was working to understand human emotions based on the facial expressions.

As these technologies are growing, they enhance the bot development, which results in faster processing of business process. This advancement can be seen in every field (i.e., military, health care department, cyber security, finance) and everywhere; these bots are changing the way of working in an organization, and they are helping humans to focus on more creative task rather than doing the same and repetitive tasks like loan clearance, document verification, data extraction, maintaining security etc.

5.8 FEATURES OF AUTOMATED BOTS

In recent years, automated bots have gained a great hype in the market, and saving cost and reducing time are not the only reasons for this hype, but also there are many reasons for it—one of the main reasons for this is precision and accuracy that simply means these bots are error-prone, and the organization don't have to think about the security and privacy of data of the users. While talking about these bots, it is must to know the ROI (return on investment), so the organization implementing bots get at least a minimum return of 30% in the first year, which grows exponentially in later years. Some of the major benefits of automation are as follows:

1. **Cost-Saving**: Bots can run 24/7 without being stopped, and these bots are only one-time investment or it can be pay as you use services they cost less. Only single bot can process the work of many employees with accuracy and in very less time. So it leads to a FTE reduction; those FTEs can be utilized at some creative and innovative places.
2. **Improved Quality**: The use of RPA process can be standardized, which can result in less error or reduce business risk. With the enhancement in the quality of process involved, the whole system is recorded and is auditable, and may adapt to changes in the process in the long run.
3. **Improved Customer Service**: More customer requests can be processed by bots without an error and provide a more optimized output.
4. **Noninvasive**: Bots don't require an integration module; they work like a normal employee of the organization.
5. **Improved Data Quality**: Using AI, data mining and analysis of more complex pattern will improve the business decisions and decrease the business risk.
6. **Scope of Data Collection**: Automation can work on any size of data with the use of deep learning, so bots can work on unstructured data.
7. **Reduce Human Access to Sensitive Data**: It's always been a problem to trust humans with sensitive data, but bots work great on this; automation technologies have a bank-grade security with high encryption techniques like AES-256 and TLS 1.2 encryption; automation technologies also maintain a clear line of separation between users, and also use granular role-based access control (RBAC).
8. **Rapid Deployment and Scalability**: Depending on the process complexity and the requirement of the organization, the bots can be developed in a short period of time and can be scaled according to the requirements.
9. **Use of Regular User Interface**: Bots use all the applications as humans like launching some applications or performing operations; if the user wants

to interact with the bots, they can interact like other applications and no backdoor things are required.

10. **Require No Complex Coding**: An RPA developer can be a business expert as RPA application comes with UI interface, and the user should have a deep understanding for business logics rather than having a knowledge for complex coding.

5.9 EFFECT OF AI AND AUTOMATION

Automation with AI is revolutionizing the whole world; these technologies have changed the processes of doing any task. These technologies have ease down all the repetitive and time-consuming tasks; people are happy to involve in some creative stuff rather than doing the same work daily, which was draining their mental health. Here are some of areas where automation plays an important work.

5.9.1 Human Resource

Automation technologies have greatly streamlined the process; they are helping to process many thousands of applications rapidly and selecting the best candidate who is well suited for the company. AI will help to select the best candidate for the organization, which in turn can result in an increase in productivity for the organization.

But for one thing, we need to understand that these bots using AI work on a large data set, and these data sets are made by humans so it needs to be make sure that these data sets should not be biased from any criteria. These bots can help to generate mass emails and are capable to maintain all the account works like creating bills, maintaining stocks etc.

5.9.2 Drones and Self-Driving Cars

This is one of the important regions for the development for the bots using AI in the field of self-driving cars; these technologies can help in the reduction of road accident and can help to move traffic smoothly, which can help in traffic reduction [4]. Surveillance drones have changed the method of security vigilance. Unmanned aerial vehicles (UAVs) are automated drones and are controlled by the remote pilot. These drones can help to carry important and required goods to the sensitive areas [5]. Even the online shopping platforms have started to move to the drone delivery system, which can help to speed up the delivery system. Amazon is one of the leading investors for making the efficient delivery drones [6].

5.9.3 Education

Automation is helping the teachers to focus more on student education rather than focusing on the repetitive tasks like giving those grades taking attendance. Exposure to these technologies will also help students to know about these technologies from the early age. These technologies can also help students who are not able to attend school by providing them with virtual classes. A great benefit can be seen in life of people who don't know to read and write or having some kinds of physical disabilities as these bots can interact with them, perform operations based upon their voice command, and provide them with the required output. It will help to provide a lifelong learning without any hindrances from our daily routine. Spaced interval learning was one of the major problems but bots help to solve this problem as they keep track of everything that individual can learn and help to remind them even a particular piece of information from all over the learning and person don't need to learn everything again. Teacher assistant, which is basically a chatbot, is able to respond to all the students' problem of any subject in a more accurate and in a faster manner. The accuracy of these bots completely depends on with how much information we train these bots [7].

5.9.4 Cybersecurity

With the increase in technology, cyberattacks have become a threat to every organization as well as to every individual. Companies have started to integrate their security system with bots and intelligence security system to detect any kind of breach in security of organization. These intelligent systems not only detect the occurrence of threat but also detect any kind of loophole that can lead to a security breach. Intelligent system not only detects the threat but also creates a system that can help to avoid any kind of attack, and let human to focus on more critical issues. But these bots can also be made to carry out some malicious activities like distributed denial of service or flooding someone's email or forum for the propagation of some message; these activities are generally done by using botnet, which is a combination of multiple bots that are used to launch a large-scale attack. For making botnet, hackers generally exploit the unpatched system, and when these bots are used, these are referred as "zombies." An effective system that participates in these kinds of attacks leads them to blacklist as IP addresses are flagged and once these bots drain down the system resources.

5.9.5 Defense Forces

With the use of technology, virtual reality forces can be made ready for many unfamiliar environments and ensure their readiness for the deadliest combat mission. These technologies have brought great changes in the field of low-light thermal image reorganization. Drones are rapidly used for the surveillance and identification of target. UAVs are used to carry out operations, and they are even capable of launching missiles and work efficiently in field of war. Intelligent bots can help to decipher the encoded message in very less time, help to deploy the solution, and predict the future threats. These systems can also identify the vehicle-borne improvised explosive devices (IEDs) or land mines. These intelligent systems can help to make the real-time tactics, which will make the soldiers to move efficiently. But automated bots and intelligent technology are used to make some more dangerous weapons and are capable of causing a greater destruction to the mankind, so many researchers are requesting to ban or limit the use of these technologies in the field of warfare.

5.9.6 Home

Intelligent bots are helping humans in every part of life: at homes, these bots do everything like controlling the lights, cooling system, and cleaning. These bots can be efficient enough to act as your personal assistant and are able to even fix appointments to doctor or saloon; they can take care for your medicine schedule and monitor your health. With the help of AI, these bots are getting efficient enough to interact according to the emotion that a person is having. In humans, these bots can take care of homes and can maintain security, and in case of misshaping, they can make a contact to local police, fire person, and any other emergency contact person. These systems use a facial recognition for the security. But privacy becomes a major concern with the increase in these intelligent systems [8].

5.9.7 Health Care

The prime areas for these bots in the field of health care are monitoring health condition, diagnosing disease, and supporting surgery. These bots can answer the frequently asked question regarding health, can send modifications for prescription refills, can make the hospital authority alert about changes in patient health condition, and can keep a medical record for further assistance. These bots can help to save lives by detecting symptoms of disease in the early stage, which can help in medical assistance [9].

Intelligent bots can be used in every field, and they are providing ease in human life but humans need to ensure that these technologies should be used in a positive way and are handled properly; if these are not handled properly, they can be hazardous to human race. These technologies can bring more destruction than a nuclear weapon.

5.10 CHALLENGES IN IMPLEMENTING AUTOMATION

5.10.1 Business Case Issues

It is difficult to convince every official and investor in the company, as it is difficult to customize RPA solution, and for the initial stage, it always seems to be costly.

5.10.2 Analysis of Process

RPA is very new technology in the market so it's important to understand which process can be automated and which can't be. It's important to understand whether by automating the process the suitable result can be achieved or not. A good team with a leader is required who have a deep understanding about the processes. The team should have very clear goals while implementing automation solutions.

5.10.3 Post-Implementation Adoption

It's a very common problem which is seen in organization that initially they don't have a team initially to handle the automation process and they are not ready to handle the upcoming challenges of automation. Automation also requires some maintenance;

one should need to maintain the output of the RPA solution. Organization need to ensure that there employees get a proper training to make use of RPA. Automation will bring a change in daily task which organizations have to ensure that there employee can cope up with that change.

5.10.4 Choosing Right Vendor

Many companies are providing the RPA solution but the organization need to choose the right vendor who can implement the solution for the organization's requirement. Using the wrong RPA solution will lead to a failure.

5.11 MYTHS OF AUTOMATED BOTS

5.11.1 Robots are Humanoid

It is a misnomer that robots are always physical, but the truth is there exists computer software that carries out the processes. These bots are capable of streamlining the process without the human intervention with the help of technologies like AI, ML, and deep networking; these automated software packages can be turned into an intelligent workflow.

5.11.2 Automation Will Replace the Human Workforce

It is one of the statements that come in everyone's mind: as the technology emerges, at some aspects it is true when it comes to RPA; many jobs are being done by automated bots. But what important here is to look at what kind of jobs—generally these are those jobs which don't want to do. It was estimated that more that 5% of all occupation can be automated and 60%–70% of all the jobs have more than 30% activities that can be automated.

But we are not looking at the fact that with the increase of these technologies, more job opportunities are being created and these are more creative jobs that require intelligence. If we see there are jobs like data scientist, social media handler, cloud architect, and many more, the problem is that people are not ready to learn new technologies and are facing the problem of unemployment.

5.11.3 Accuracy

We need to understand that these are just a set of commands which completely depend on how the accurate developer was while designing the bot. So if wrong commands have been given by the developer or bots have been trained less or trained with wrong data set, it's never going to produce the right result.

5.11.4 Expensive

Automation is cheap when we look to a bigger picture; it just costs one-third of an employee wages. And it's not because those bots can work 24/7 without break, but it's because of the reason that 1 minute of bot work is equivalent to 15 minutes of work by an FTE.

5.11.5 Internal Environment of Organization

Before using the automation in the organization, first focus on the employees and the processes. It is important to see that the company does have enough resources to implement automation, and have to perform the feasibility study for all the processes we want to automate and to find the suitability of operation.

5.11.6 Robots Can Be Left Unattended

This is not true; some kind of control is always required for scheduling, running the bots, handling exception, processing output, and many other things.

This also proves that bots can't replace humans; they will always work on the control point, as they have only that much of intelligence which they have been coded by the developer.

There are still thousands of questions that are left unanswered when it comes to AI and automation—some of them are as follows:

1. What moral values should we embed in them so that they can act and coexist with human feeling and how we are going to teach bots to act according to our interest.
2. How we can teach machine that what is ethical, like it may happen that there is way which is efficient but not ethical; moreover, these change over time, culture, and the requirement. Likewise, there is white hat and black hat in hacking.

   The point is to make things efficient, but if they don't act in a natural way or as per the demand of situation, then it could be a problem.
3. The automated system records everything; it keeps track our every movement so privacy can be a major concern.
4. AI algorithms learn from data set training; some of the results are complex to understand and some of the results are not even understood by the designer of the algorithm, which mean the system is not going to opaque, and as the technology will become more efficient, these algorithms can become a black box for humans.
5. Who will be left accountable if the whole automated system gets crash someday, or what can be the problem that can be caused if someone bypasses and changes some code of the system.
6. These systems can be learned by the data sets, but the data sets are created by humans, what if these data sets are biased, and what kind of result they can produce. How these systems can overcome from the biased data or is there any way to identify that these data are biased [10].
7. The whole social media is being controlled by the automated software, what if they start spreading some biased information, which can be very

manipulative for the society and even lead to riots, and war between countries, or bring political changes.

8. One major question is what if these systems become much intelligent that they start developing themselves and start rewriting their own code, then there will be no human control over these systems and these systems will become completely opaque.

Questions like these make scientist to think whether the automated system will really make the human life better.

5.12 PLATFORM USED FOR IMPLEMENTATION

Some basic technologies used for the implementation of automation and AI are shown in Figure 5.5.

5.12.1 PYTHON

Apart from the reason that it is easy to learn, it is rich in libraries and has many libraries for data mining and analysis, neural network libraries, and many others.

5.12.2 TENSOR FLOW

This is also a python library developed by Google for neural network, deep learning, and other regression algorithms.

5.12.3 R

It is a programming language and can be used to automate processes that are involved in data processing and manipulation.

5.12.4 SCIKIT-LEARN

This is also a library of python that is used for data mining, data analysis, and ML algorithms.

5.12.5 AUTOMATION ANYWHERE

This is an RPA tool that helps to automate the business processes. It is a UI-based system and requires less coding knowledge to use and provide run-and-playback feature and provide cognitive capability.

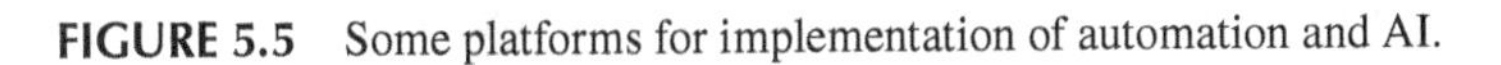

FIGURE 5.5 Some platforms for implementation of automation and AI.

5.12.6 UiPath

This is similar to automation anywhere but provides a visual design and works efficiently in BPO operation.

5.13 CONCLUSION

With the great advancement in technology, AI together with automation is revolutionizing the whole world, and there are only some aspects of these technologies, but a lot is yet more to come. These technologies have brought some great ease to the hectic life of people. Automation is becoming a part of every individual life. So, it is about how the individual is going to use these technologies as these technologies can also bring some very bad effects to the life of individual, which need to be handled in a way to bring out the maximum advantage. These technologies are taking up all the works which humans hate to do the entire repetitive task, and there is a lot of new job opportunities that are creative and are of great interest. Both automation and AI are not single technologies, and they are the combination of many technologies like NLP, reinforcement learning, ML, and many others; a lot of work can be done in these fields. These technologies are going to be the future of humans.

REFERENCES

1. Khurana D., Koli A., Khatter K., Singh S. (2017). Natural language processing: State of the art, current trends and challenges. arXiv preprint arXiv:1708.05148.
2. Suri V.K., Elia M., van Hillegersberg J. (2017). Software bots – The next frontier for shared services and functional excellence. In: Oshri I., Kotlarsky J., Willcocks L. (eds) *Global Sourcing of Digital Services: Micro and Macro Perspectives. Global Sourcing 2017.* Lecture Notes in Business Information Processing, vol. 306, pp. 81–94. Springer, Cham.
3. Lamberton C., Brigo D., Hoy D. (2017). Impact of robotics, RPA and AI on the insurance industry: Challenges and opportunities (November 29, 2017). *Journal of Financial Perspectives*, vol. 4, no. 1. Available at SSRN: https://ssrn.com/abstract=3079495.
4. Rogers C. (2015). Google sees self-driving cars on road within five years. *Wall Street Journal.* http://www.wsj.com/articles/google-sees-self-drive-car-on638road-within-five-years-1421267677.
5. Floreano D., Wood R.J. (2015). Science, technology and the future of small autonomous drones. *Nature*, vol. 521, pp. 460–466.
6. Joshi, D. (2017). Exploring the latest drone technology for commercial, industrial and military drone uses. *Business Insider.* Retrieved from https://www.businessinsider.com/drone-technology-uses-2017-7.
7. Lasso-Rodríguez R., Gil Herrera R. (2019). Robotic process automation applied to education: A new kind of robot teacher?. pp. 2531–2540. doi: 10.21125/iceri.2019.0669.
8. Asadullah M., Raza A. (2016). An overview of home automation systems. doi: 10.1109/ICRAI.2016.7791223.
9. Hamet P., Tremblay J. (2017). Artificial intelligence in medicine. *Metabolism: Clinical and Experimental*, vol. 69, pp. S36–S40. doi: 10.1016/j.metabol.2017.01.011, PMID: 28126242.
10. Snow, J. (2017). New research aims to solve the problem of AI bias in 'black box' algorithms. *MIT Technology Review.* Retrieved from https://www.technologyreview.com/s/609338/new-research-aims-to-solve-the-problem-of-ai-bias-in-black-box-algorithms/.

6 Artificial Intelligence
An Emerging Approach in Healthcare

Yash Tyagi and Pardeep Kumar Sharma
Lovely Professional University

CONTENTS

6.1 INTRODUCTION

In terms of technology, artificial intelligence (AI) (man-made brainpower), also referred as machine knowledge, is the understanding displayed by the machines, rather than the characteristic knowledge displayed by the humans. Driving AI reading material characterizes the area as the "wise operators" investigation: any gadget that optimizes according to its ambiance and completes its task effectively to accomplish the desired objectives [1]. Colloquially, the term "man-made reasoning" is regularly used to portray devices (PCs) that represent "intellectual" works which individuals link with humanoid psyche such as "learning" and "issue solving" [2].

AI's history began in the days of past eras, with myths, tales, and bits of gossips about counterfeit creatures endowed with intelligence or intellect by skilled workers. The conventional researchers planted the seeds of the present-day AI who endeavored to depict the procedure of human intuition as the mechanical control of images. This work brought about the development of programmable and digitalized PCs in the 1940s, which are machine-dependent on numerical thinking's theoretical embodiment. This gadget and the thoughts behind it propelled a bunch of researchers to start truly talking about the chance of constructing an automated brain. The concept of AI research originated in 1956 at Dartmouth College [3], where John McCarthy authored the term "man-made brainpower" to acknowledge the field as an entirely different area of study and to break the influence of cyberneticist Norbert

Wiener [4]. Attendees of the first conference to discuss the concept of "man-made brainpower"—Herbert A. Simon (CMU), Arthur Samuel (IBM), Marvin Minsky (MIT), Allen Newell (CMU), and John McCarthy (MIT)—became founders and organizers to investigate AI.

In the field of healthcare too at that time, people had certain preconceived ideas but believed that one day AI would revolutionize the medical world. These preconceived thoughts no longer remained ideas as in the 1960s and 1970s pieces of research that produced the first expert system (problem-solving program) "Dendral," which mechanized the dynamic procedure and critical-thinking conduct of organic chemists. This master framework changed the point of view of the researchers, and AI was viewed as an innovation that could refurbish several clinical aspects and could also ease out the execution of tasks within manufactures, consumers, and the pharmaceutical organizations. Numerous researches suggest that AI can execute certain healthcare tasks just as or better than humans, including diagnoses of diseases. Calculations already beat radiologists today to spot threatening tumors and guide scientists on how to build companions for exorbitant preliminaries in clinical practice. However, we have to agree to the fact that it's still a long way to go, when AI would be completely able to replace people practicing in the healthcare field. Further in this chapter, we would be detailing on both the potential that offered for mechanizing healthcare components and the segment of the limitations/challenges faced by the rapid implementation of AI in medicinal services [5].

6.2 SCOPE & RELEVANCE OF VARIOUS TYPES OF AI IN HEALTHCARE

AI in the health industry is essentially the utilization of multifaceted programming and calculations to imitate human intellect in studying, interpreting, and appreciating complicated data about clinical and medical services. In general, simulated intellect is the ability of computer to accomplish the desired targets with no human involvement and utilizing the algorithms.

The distinguishing factor of AI innovations from the modern healthcare applications serving as the normal advancement in medical services is its ability to collect statistics, channelize it, and deliver the required results to the end user or the medical practitioners [6].

Before AI structures can be submitted for the application in the healthcare, they should be equipped with the statistical data and the other figures that are generated from medical activities such as diagnosis, screening, treatment etc., with the intention of being able to learn relation between subject highlights, comparative subject gathering, and premium outcomes.

However, this clinical information is not confined on a regular basis by the type of the clinical notes, physical evaluations, socioeconomics, clinical laboratory and images, and electronic chronicles from clinical gadgets [5].

Essentially, we must realize that AI is not an invention, but rather an amalgamation of innovations. The vast majority of such advancements are of prompt importance to the field of medical services, but the errands and specificity of the tasks that are accomplished by exploiting the concepts of AI vary generally. Some specific AI developments of great significance to the healthcare are listed and briefed below.

- Machine Learning—Deep Learning & Neural Network

  AI is a realistic method based on "learning" information by planning models. It is an expansive strategy at the central core of innumerable ways of dealing with AI, and numerous renditions of it exist. The most commonly accepted application of customary AI human services is envisaging medicines with precision. This activity is entirely centered on what medical standards mostly prevail in a patient based on different symptoms of the subjects and the medication setting [7].

  The mainstream applications of AI and accurate medication include the preparation of datasets for which the variable outcomes (e.g., initiation of ailment) are previously known. And it is called the administration learning.

  An increasingly unpredictable form of AI is the neural system—an innovation that has been accessible from the1960s and has been entrenched in the medicinal services examined for quite a few years. It has been applied in the healthcare for deciding whether a patient is likely to get a particular disease [8]. It is correlated with the manner in which neurons interpret signals, but its resemblance to the functionality of the cerebrum is usually feeble. Most of the mind-boggling types of AI comprise the profound learning or neural system models that foresee results with a variety of related factors. There are a large number of hidden layers incorporated in such models, which are discovered out by the quicker handling of the present designs, preparing units, and cloud structures. A characteristic use of profound learning in human services is the acknowledgment of possibly destructive injuries in radiology pictures [9]. The use of deep learning has been pragmatic in radiomics or in the recognition of clinically important highlights in imaging information, which aren't visualized by the naked eye [10]. Both radiomics and profound learning found use in the oncology-based picture investigation. Their blend seems to guarantee more prominent precision in finding than the past age of mechanized devices for picture investigation, known as PC helped location or CAD [11]. Profound learning is likewise progressively utilized for discourse acknowledgment and, all things considered, is a type of normal language handling (NLP), depicted beneath. In contrast to prior types of measurable examination, each component of profound learning model has been of least importance to a human spectator. Thus, the elucidation of the model's results might be troublesome or problematic for the resource person.
- Robotic Process Automation

  This technology carries out advanced structured tasks for governing purposes; i.e., it includes data framework, as if it was a human client keeping content or regulations. Robotic process automation is economical, easy to program, and straightforward in their operations, contrasting with various types of AI. Robotization of mechanical procedures (RPA) doesn't imply the utilization of robots, which are simply the computer systems with definite codes on service. It depends upon the amalgamation of the work processes, market procedures, and the integration of the "external coating" with the data structures to serve as a system's client having a partial intellect.

RPA is used in medical services for dull errands like refreshing medical data and earlier approval. When paired with various advancements such as image recognition, RPA can cite information from, for example, faxed imageries in order to incorporate them into the value-based systems [12]. However, portraying these innovations as distinct entities would be nothing more than a mirage, because gradually we find that all these technologies are joined and coordinated; robots are now being incorporated with brains having a simulated intellect; picture acknowledgment has been linked to RPA. Maybe, later on, these advancements will be blended to a point where composite arrangements will be more probable or attainable.

- Rule-Based Expert System

  During the 1980s, master structures based on the collection of "if-then" rules were the predominant expertise for AI and broadly found the use in industries at that time and even later. These have been commonly applied in the medical services for purposes of "clinical decision support (CDS)" in the past few decades, and even today they are extensively used [11].

  Various electronic health record (EHR) suppliers incorporate numerous regulations in their frameworks. Extensive utilization of professionals and data designers is involved in expert systems to construct a consortium of regulations in a space of explicit data.

  They work excellently to a small degree and are transparent. But if the amount of regulations is big (>1000), and the core values begin to compete with each other, they tend to detach. Additionally, modifying the rules can be difficult and tedious if the information area changes. Methodologies relying on the information, algorithms, and AI calculations are slowly replacing them in medical services.

- Natural Language Processing

  Since the 1950s, understanding human language and decrypting it in terms of algorithms has remained the primary goal of many AI scientists. The domain of NLP incorporates various works and uses like content examination, discourse acknowledgment, and interpretation. There are two elementary ways to address it: the factual and the semantic NLP. Measurable NLP is related to machine learning (specifically deep learning neural systems) and has added to an ongoing enhancement in the exactness of recognition. It needs a large "corpus" or community of language to learn from.

  NLP's primary uses in medical services include the development, interpretation, organization of clinical data; receive ready reports (e.g., analysis of radiology); decode a consistent collaboration; and guide conversational AI.

- Physical Robots

  The physical robots have become noteworthy by now, as an enormous number of advanced robots greater than 200,000 are positioned around the world annually. They carry out precharacterized activities like repositioning, elating, fusing, or amassing items in industrial plants, granaries and storage rooms, and facilitating hospitals with the required items. As of late, robots have become more and more community-oriented with humans, and are all the more handily equipped by pushing them around the perfect errand. They are also becoming progressively smart, as other AI capabilities are being installed in their "cerebrums" (actually in their chips or operating framework).

With time, it's likely that physical robots would get more advanced and modernized as the other sectors of simulated intellect contract the stairs of development. Surgical robots gave "superpowers" to the specialists, when official permission was granted to their use in the United States in 2000, improving the capability to visualize and reducing the probabilities of errors. However, all the critical decisions while performing the surgeries are still dependent on the cerebrum of the specialists [13]. Fields in surgery wherein this modernized robotic technology is being incorporated are gynecology, prostate, and head-and-neck medical procedures.

Lately, it has been seen that simulated intellect strategies have set remarkable milestones and have proved out to be an extraordinary invention across medicinal services. Now and then, a fueling conversation related to whether AI researchers can replace the doctors and surgeons in the healthcare industry in the long run still remains a major topic of concern. However, we agree to it that the probability of supplanting health professionals with modern mechanization isn't possible till a reasonable timeframe, but yes we even can't neglect the fact that this artificial knowledge Incorporated in the health industry enhances the diagnostic capability of the medical practitioners & has even surpassed the clinician's ability in certain utilitarian zones of health science, like radiology. Extending availability of human administration data and the quick headway of enormous data-demonstrative procedures have made possible the progressing viable employments of AI in restorative administrations. Guided by the relevant clinical requests, mind-blowing AI systems can open clinically huge information concealed in the colossal proportion of data, which subsequently can support the clinical decision-making [14].

Figure 6.1 delineates the guide from the clinical information generation to normal language handling information advancement, to AI information examination, to clinical dynamic. [15].

6.3 AI'S TIMELINE IN HEALTHCARE

At the point when a large number of us hear the expression "man-made brainpower" (AI), we envision robots carrying out our responsibilities, rendering individuals out of date. What's more, since AI-driven PCs are modified to settle on choices with a minimal human mediation, some miracle if machines will before long settle on the troublesome decisions we currently depend on our primary care physicians. As indicated by David B. Agus, it's essential to differentiate AI from Sci-Fi, because AI is already here, and it's gradually evolving medication. As opposed to mechanical autonomy, AI in medicinal services, for the most part, suggests the health professionals and emergency clinicians with some remarkable data figures that enhance the

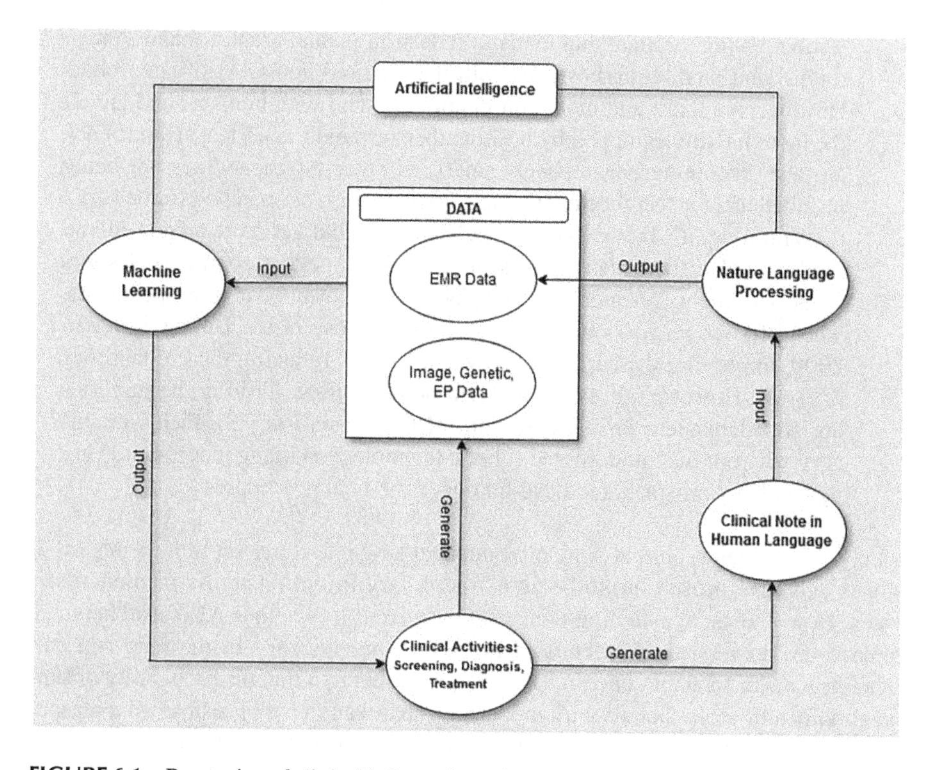

FIGURE 6.1 Processing of clinical information using AI. (Taken from reference number [15].) EMR: electronic medical record; EP: Electrophysiology

quality of treatment provided to a patient, within a short span of time. This incorporates the test results, analyzation of the reports, enhanced survival rates, and the rate of well-being across the globe.

All this that we have discussed right above is not a one-day miracle; this is the result of a lot of time and brainstorming of the medical experts and scientists that today we are thinking of such advancements in technology and replacing human brain with an electronic one [16].

It all started with the research during the 1960s and 1970s that created the primary critical-thinking project, or master framework, called "Dendral" [17]. Though it intended its use in the natural science, it gave the premise to a resulting framework "MYCIN" [18], regarded as the foremost application of the simulated intellect in the medicinal industry. Conversely, "MYCIN" and different frameworks, for example, INTERNIST-1 [19] and CASNET, didn't accomplish the repetitive use by the professionals [20].

The years of 1980–1990s saw the development of the microsystems/computers and modern forces of framework accessibility. It was the time, when after an affirmation by scholars and specialists of AI, it was expected that simulated structures in social protection must suit the absence of clinical and medical experts [21]. Several theories above the scope of this chapter help in judicious acceptance of artificial structures in human administrations [22].

Clinical and mechanical progressions over the time span of 50 years [23] have sanctioned the growth of medicinal services that allied utilization of AI to include:

- Quicker data collection and its flow that was being enhanced gradually [22].
- Enhancing the NLP and system's visualization, authorizing machines to duplicate human interpreting skills [21].
- Upgrading the precision in robot-assisted surgeries [23].
- Development of genomic sequencing databases [19].
- Upgrades in profound learning strategies and information signs in uncommon ailments.
- Far-reaching usage of electronic well-being record systems [20].

Table 6.1 depicts the brief historical timeline of AI in healthcare [16].

TABLE 6.1
Brief Historical Timeline of AI in Healthcare

Time Period	Event
Mid-1950s	A regular objective of early endeavors at clinical master frameworks is to supplant the doctor with a Greek prophet model of clinical dynamic. To make a "specialist in a container," fit for questioning the doctor or clinical professional regarding the manifestation of a patient, and to produce a conclusion is the purpose of this line of belief
Late 1960s	Early master frameworks started impressive energy in the field of medication, and the late 1960s brought about an elevated degree of desire
1970	In medication, a persuasive paper was published by Dr. William B. Schwartz in the *New England Journal of Medicine* about AI. Accordingly, numerous researchers are pulled in to examine the utilization of software engineering in medication
1970	A joint establishment was set by Harvard Medical School and MIT-related training, research, and administration to cultivate the improvement of well-being. Health, Sciences, and Technology (HST) were named the new division. Among numerous projects, HST has offered the preparation in clinical informatics, which is a field firmly identified with AI in medication
1975	Improvement of MYCIN master framework, a standard-based program for the conclusion of bacterial diseases in the blood (Stanford)
1975	Advancement of the INTERNIST master framework, an analytic guide that consolidates an enormous database of illness/sign relationship with methods for issue definition (Pittsburgh)
1976	Advancement of CASNET (Causal-Associated Network) master framework, which utilizes physiological models for the finding and treatment of eye malady (Rutgers Research Resource)
1977	Advancement of PUFF master framework, for programmed translation of pneumonic capacity tests (Pacific Presbyterian Medical Center, San Francisco) In its business structure, the PUFF framework has been sold to several international destinations that are still being used today

(Continued)

TABLE 6.1 (*Continued*)
Brief Historical Timeline of AI in Healthcare

Time Period	Event
Late 1970s	Advancement of PRESENT-ILLNESS master framework, to analyze kidney ailments (MIT) It utilizes a PC model, which is more complex than the standard-based frameworks used in different ventures, an "outline-based" model where an edge-like structure inside the program speaks to living things
1979	Foundation of the American Association for Artificial Intelligence (AAAI)
1981	Improvement of the ABEL master framework, a program that utilizes staggered pathophysiologic models for the conclusion of corrosive base and electrolyte issue (MIT)
Early 1980s	Simulated intelligence in medication is to a great extent US-based research network. Work starts out of numerous grounds, including MIT, Pittsburgh, Stanford, and Rutgers
1983	Improvement of a PDS master framework (CMU)
1983	Improvement of the MED1 master framework (Kaiserslautern)
1984	Clancey and Shortliffe give the accompanying meaning of clinical AI: "Clinical computerized reasoning is worried about the development of AI programs that perform determination and make treatment proposals. Not at all like clinical applications dependent on other programming strategies, for example, have factual and probabilistic techniques, clinical AI programs depended on emblematic models of malady elements and their relationship to tolerant variables and clinical signs" Much has changed from that point forward, and presently, this concept would be viewed as tight in degree and vision. Currently, the significance of finding as an errand needs PC assistance in regular clinical circumstances gets substantially less accentuation
1985	Development of MED2 expert system (Kaiserslautern)
Mid-1980s	Foundation of an organization called the fifth Generation Project in Japan, whose goal was to facilitate the communication among medical AI scientists around the world
Mid-1980s	The advancement of master frameworks experienced a couple of downsides as of now: • Some master frameworks couldn't work just as the specialists who provided them with information • Most of the master frameworks must be run on (expensive) LISP machines • The LISP machines couldn't be associated with a system These constraints upset the improvement of master frameworks as business applications at that point
1986	Foundation of European Society for Artificial Intelligence in Medicine (AIME) in Europe
1989	Improvement of MUNIN master framework for diagnosing neuromuscular disarrangement
Early 1990s	The disappointment of scientists to introduce good frameworks brought about diminished subsidizing from the legislature and financial speculators in the mid-1990s. Man-made intelligence researchers regularly allude to the time of 1987–1994 as the "artificial intelligence winter"
1991	Improvement of PEIRS (Pathology Expert Interpretative Reporting System) master framework, to produce pathology reports. PEIRS covered an assortment of pathology measures, with a general indicative exactness of about 95%

(*Continued*)

TABLE 6.1 (*Continued*)
Brief Historical Timeline of AI in Healthcare

Time Period	Event
1993	Improvement of the GermWatcher AI research center framework. This framework checks for emergency clinic-gained (nosocomial) diseases, which speak to a huge reason for delayed inpatient days and extra medical clinic charges Microbiology culture information from the medical clinic's laboratory framework is observed by GermWatcher, utilizing a standard base containing a blend of national rules and nearby emergency clinic disease control approach.
Mid-1990s	Research in clinical AI changes its concentration to new territories, including: • Getting more information on the web building a better data foundation • Using AI methods to decide • Learning and growing better simple to utilize programs reasonable for well-being experts A focal piece of every one of these activities was the production of electronic medical records (EMR), which fills in as the focal clinical store of data on persistent consideration
1997	The (American) National Library of Medicine grants agreements to an assortment of human services associations of the nation over to examine inventive employments of the national data framework for social insurance, including telemedicine and data sharing
Near Future	Completing the human genome project will certainly lead to the following AI applications in medicine: • The collection of the information generated through the use of AI systems in hospitals around the world will be linked, enabling them to share all the information. The AI system will aggregate this information utilizing distributed computing and subsequent analysis which will draw inferences from data mining applications that will produce patterns based on this collected data • To determine the important patterns, expert systems and neural networks will be used • To ensure a system that the system should continue to learn, genetic algorithms should be used The analysis result will be fed back to the AI inference engines of individual hospitals to allow their AI software to analyze data of each patient regarding the present patterns of surgery complications, diseases, medical complications with certain types of the genome, etc.

Data taken from reference number [16].

6.4 IMPLEMENTATION OF AI CONCEPTS IN THE MEDICAL WORLD

From the most punctual crossroads in the cutting-edge history of the PC, researchers have longed for making an "electronic cerebrum." Of all the cutting-edge mechanical missions, this hunt to make shrewd and deceitfully intellectual computer frameworks has been one among the supremely driven and, as anyone might expect, questionable missions. It additionally appears that at an opportune time, researchers and the specialists were spellbound by the potential such an innovation may have in medication. With astute computers ready to store and process tremendous stocks of information, the expectation was that they would become flawless "medical specialists in a crate," helping or outperforming clinicians with tasks like diagnosis.

In the past years, AI in medication has gradually picked up its ubiquity. Clinical AI frameworks, and the advances for building up these frameworks, have been demonstrated in an immense range of standard clinical applications. However, man-made consciousness has made the hectic tasks easier and has reduced the expenditure and time consumption on those tasks saving the mankind from wasting a whole lot of energy, money, and labor. It has also enhanced the relationship between the diseased person, health professionals, and medical overseers. The simulated intellect—one of the recent trends of the world and the most researched concept—was valued at about $ 600 million of every 2014 and is expected to grasp a value of $ 150 billion by 2026 [24].

Computer-based intellect has interminable applications in healthcare amenities. AI has been aiding the medical amenities irrespective of the domain it is utilized for. Whether the connections between the genes/hereditary codes are to be studied, or the ways to control mechanized robots and enhance medical expertise are concerned, AI has proven to be a boon to the healthcare industry. Now moving from the past to the present, the transformations that AI has brought in healthcare can be summarized by analyzing the developed technologies and AI-based applications that are aiding the healthcare professionals. Few domains and applications where the professionals are using the AI excessively are enlisted below.

- Efficient Employment of AI in Diagnosis & Reducing Human Errors

  In 2015, misdiagnosing sickness and a clinical blunder represented 10.0% of all US mortalities. Considering that, the guarantee of refining the symptomatic procedure is one of AI techniques most exciting and vital use in the public sector.

  Extremely less clinical data and more pressure on healthcare professionals of patients can cause blunders in diagnosis. But while using AI as it is unaffected by those factors, it can determine the chances of mortality faster as compared to the determination by the doctors. In a study, for example, man-made brainpower model utilizing calculations and profound wisdom determined breast cancer faster than 11 clinicians. The various examples/different ways AI is diminishing mistakes and sparing lives are enlisted below.

  - **PathAI Made Cancer Diagnosis More Accurate with AI**: PathAI is making novelties in the simulated intellect to help clinicians mark increasingly precise decisions. The administration's existing purposes are to diminish mistakes in malignant growth analysis and generate tactics for the personalized clinical management.
  - **Enlitic Streamlined Radiology Diagnosis via AI Deep Learning**: To smoothen the radiology analysis, Enlitic has grown profound learning therapeutic gadgets. The corporations' profound learning stage assists and allows the clinical professionals and specialists to better comprehend a person's/subject's continuous necessities by breaking down

the unstructured laboratory info, incorporating test results, radiologic images, genomic studies, and subjects' past medical reports of any kind.

- **Early-Stage Diagnosis of Deadly Blood Diseases**: Man-made reasoning has been utilized at "Schooling Hospital of Harvard University" to analyze possibly dangerous blood infections at a beginning time. There the experts are also utilizing the advanced and enhanced AI-magnifying lenses to filter for hurtful microscopic organisms (e.g., *E. coli* and *staphylococcus*) in the blood samples faster than the preexisting physical examination. After visualizing more than 25k image samples of blood tests, the machines are learnt to scan for microscopic organisms. And with such learning, the machines could recognize and anticipate destructive minute bacteria and pathogens present in the body fluid with 95.0% precision.
- **AI-Based Symptom Checker**: Buoy Health—a simulated intellect-based application—utilizes calculations to analyze and treat sickness. In this checker, a chatbot tunes into a subject's indications and well-being fears, and assists the subject as per its findings and queries raised. Buoy's simulated intellect is extensively used across the globe by hospitals and medical industries—one of them is Schooling Hospital of Harvard University. This innovation of the multinational corporation has led to the speedier recovery of the diseased person.

- **AI Working as an Assistant in Radiology**: The machines having false intellect help the radiologists by studying the images and providing them with facts that help the radiologists make clinical decisions.
- Use of AI in Medicine Development in the Pharma World

The advancement in the medication industry is impeded by soaring improvement expenses and research that takes a great many humanoid minutes. It budgets around $ 2.6 billion for each medicine to pass through experimental trials—out of which, only 10.0% of the medicines are commendably carried in the market. But now the pharmaceutical multinational corporations with progressions in innovation have hastily payed attention towards the proficiency, exactness, and facts that the simulated intellect can provide with.

The most premium artificial intellect discovery for the improvement of medicates originated in 2007 when scientists entrusted an automaton named "Adam" that examined the features of yeast. It battered billions of info that focuses on uncluttered catalogues to hypothesize about the elements of 19 qualities exclusive to toadstools, anticipating nine innovative and precise speculations.

- **Using Deep Learning for Targeted Treatment**: The essential objective of BenevolentAI lies in finding a correct cure to the right person at the opportune period by utilizing computerized reasoning to deliver a superior objective choice and give beforehand unfamiliar experiences through profound learning.
- **Polymorph Prediction in Drug Discovery**: Consolidating fake wisdom, quantum material science, and concoction of medicinal properties of the little particle contender, XtalPi's ID4 stage predicts the enhancement in the modernized medical plans. Moreover, the company affirms

its innovation of predicting the multifaceted subatomic structure of the drug particle within days instead of weeks and months.

- **Employing AI in Biopharmaceutical Development**: Drug discovery in the domain of neuroscience and immune-oncology has gradually contacted with the concepts of artificial intellect. BioXcel Therapeutics' medication re-development program utilizes AI to discover new applications for existing medications or to recognize new patients. BioXcel Therapeutics' work in AI-based medication improvement was named as one of the "Most Innovative Healthcare AI Developments of 2019."
- **AI Functioning in Clinical Trials**: Atomwise utilizes AI to handle a portion of the present most genuine infections, including Ebola and numerous sclerosis. The organization's neural system, AtomNet, predicts bioactivity and recognizes the persistent attributes for clinical preliminaries. Atomwise's AI innovation screens somewhere in the range of 10 and 20 million genetic tests everyday and can convey results multiple times quicker than conventional pharmaceutical organizations.
- **Fighting the Rare Diseases with AI**: BERG is a clinical platform, simulated intellect-grounded biotech stage that maps illnesses to quicken the disclosure and advancement of drugs. By consolidating its "Inquisitive Biology" approach with conventional R&D, BERG can grow increasingly hearty item up-and-comers that battle uncommon maladies. BERG as of late introduced its discoveries in the treatment of Parkinson's disease by utilizing simulated knowledge for discovering connections between the unclear synthetic compounds in individual's body.
- **Using Artificial Intellect to Discover Better Candidates for Developmental Drugs**: Profound Genomics' AI stage assists scientists with discovering the possibility for formative medications identified with neuromuscular and neurodegenerative clutters. Profound Genomics is likewise taking a shot at "Undertaking Saturn," which can examine more than 69 billion of distinctive cell mixes and furnish technologists with input.

- Artificial Intellect Streamlining User Experience

In the health facilities, time is money. Productively channelizing the inflow and outflow of diseased person permits emergency centers, facilities, and doctors to treat more patients every day. An enormous crowd of 35 million in 2016 with various afflictions, protection inclusion, and conditions was witnessed by specialists in the US emergency clinics. Investigation in 2016 incorporating 35k doctor surveys uncovered that the absence of client assistance, disarray over administrative work, and unresponsiveness of specialists at time of emergency became the reason of 96.0% of client protests.

Now with advancements in the artificial intellect and its heavy engagement in the healthcare field, the user experience is streamlining and is helping the healthcare professionals and other staff to process millions, if not billions of cases within a short span of time with precision. Few instances of how AI is helping healthcare departments in managing the patients flow are as follows:

- **Healthcare Plans Customized with AI**: The historic collaboration of the Cleveland Clinic with the IT tycoon IBM for the advancements in the man-made brain power has smoothened the user experience and has allowed the company to customize its medicinal services to a greater extent.
- **AI Recommending the Need for Check-Up**: Babylon utilizes AI to give customized and intelligent medical assistance, including face-to-face meetings with specialists whenever required. The organization's AI-fueled chatbot streamlines the audit of a patient's side effects, at that point suggests either a virtual registration or an in-person appointment with a medical staff proficient. Babylon and Canada's Telus Health collaborated to build up a Canada-explicit AI application that filters a patient's study answers and connects them with appropriate health advisors or experts for further guidance and diagnosis.
- **Eliminating the Delay in Treatment using AI**: Qventus based on artificial intellect is a programming stage that fathoms operational difficulties, including those identified with crisis rooms and patient well-being. The organization's computerized stage organizes persistent ailment/injury, tracks emergency clinic holding-up times, and can even outline the quickest rescue vehicle courses.

  CB Insights named Qventus is one of its 100 most innovative AI startups for 2019 dependent on the organization's work in mechanizing and organizing quiet well-being.
- **Utilizing Machine Learning for a Better Patient Journey**: Cloud MedX utilizes AI to produce experiences for improving patient excursions all through the human services framework.

  The organization's innovation enables emergency clinics and facilities to oversee understanding information, clinical history, and installment data by utilizing a prescient examination to mediate at the basic points in the patient consideration experience. Medicinal services suppliers can utilize these bits of knowledge to proficiently move patients through the framework with number of the customary disarray.
- **Robotizing Healthcare's Most Repetitive Processes**: Olive's artificial intellect-based platform is intended to robotize the medical field's various monotonous assignments, opening up directors to deal with more significant level ones. Here, AI has helped the staff members to concentrate on offering healthier client support, by robotizing everything from information relocation to studying the basic client history and booking appointments, as and when required.

  Olive's artificial intellect-based platform effectively incorporates inside an emergency clinic's current programming and apparatuses, wiping out the requirement for expensive reconciliations or personal times.
- **AI Prioritizes Hospital Motion to Assist the Diseased Persons**: Johns Hopkins Hospital as of late reported its collaboration to GE to utilize prescient AI procedures to enhance the productivity of subject's operative stream. A team, expanded with computerized reasoning,

immediately organized an emergency clinic movement to support all the diseased persons. A 60.0% enhancement in its capacity to concede clients and 21.0% expansion in understanding releases before afternoon, has brought about a quicker, progressively positive patient experience, after the company actualized the program.

- Managing & Taking out Therapeutic Facts with Artificial Intellect

Health services are broadly viewed as one of the enormous information wildernesses to be domesticated. Profoundly important data can sometime be misplaced amid the timberland of tons of information focuses, trailing the business around $ 100 billion every year. The ease to associate significant information is allowing better advancement of new medications, deterrent medication, and appropriate determination. Numerous social insurances are going to man-made consciousness as an approach to stop the information discharging. The innovation separates information storehouses and interfaces in no time which previously took a long time to process.

Beneath we have some instances of artificial intellect-based organizations serving the medicinal services commerce to remain awash in an expanse of information.

- Tempus has been utilizing artificial intellect to filter through the world's biggest assortment of medical and atomic information to customize social insurance medicines. The organization is creating AI instruments that gather and break down information in everything from hereditary sequencing to picture acknowledgment, which can give doctors better bits of knowledge into medicines and fixes. Tempus is as of now utilizing its AI-driven information to handle disease research and treatment.
- IBM's Watson is serving human services experts outfit their information to upgrade medical clinic productivity, better communication with patients, and enhance treatment. Watson is right now smearing its abilities to the lot from creating customized well-being plans to deciphering hereditary testing outcomes and getting timely indications of sickness.
- KenSci consolidates huge information and man-made consciousness to anticipate clinical, money-related, and operative hazards by taking information from current sources to predict everything from who may become ill to what's pouring up an emergency clinic's human services costs.
- H2O.ai's AI breaks down information all through a medicinal services framework to mine, robotize, and anticipate forms. It's being utilized to foresee intensive care units moves, enhance laboratory work processes, and even identify a subject's danger of medical clinic procured diseases. Utilizing the organization's man-made reasoning to mine well-being information, clinics can foresee and recognize sepsis, which at last diminishes passing rates.
- Proscia is an advanced pathology stage that utilizes AI to identify designs in disease cells. The company's product allows pathology laboratories to eliminate data bottlenecks and uses AI-powered image analysis to connect data points that aid in the detection and treatment of malignant growth.

  - Nowadays emergency clinics are utilizing Google's DeepMind Health artificial intellect programming everywhere around the world to assist transferring diseased persons from testing to cure all the more productively. It informs the specialists when a person's well-being breaks down and can even assist in the finding of diseases by looking over its gigantic dataset for similar indications. By gathering manifestations of a person and contributing them to the DeepMind stage, specialists can analyze rapidly and feasibly.
- AI Robots Assisting in the Critical Life-Saving Surgeries

  Ubiquity in robot-helped medical procedures is soaring. Medical clinics are utilizing robots to help including insignificantly intrusive methodology to open-heart medical surgical procedure. As per the Mayo Clinic, automatons assist specialists in performing complex techniques with accuracy, adaptability, and control that go past human abilities.

  Robots furnished with cameras, mechanical arms, and careful instruments increase the experience, expertise, and information on specialists to make another sort of medical procedure. Specialists regulate the machine-driven arms while situated at a PC comfort, while the automaton gives the specialist a three-dimensional, amplified perspective on the careful spot that specialists couldn't get from depending on their naked eyes. The specialist at that point leads other colleagues who work intimately with the automaton through the whole activity. Robot-helped medical procedures have prompted less medical procedure-related difficulties, less torment, and faster recuperation time. Investigating few examples of the same are described below.

  - Vicarious Surgical joins augmented reality with AI-empowered robots so specialists can perform insignificantly obtrusive tasks. Utilizing the organization's innovation, specialists can recoil and investigate within a patient's body in substantially more detail.
  - Specialists utilize the Mazor Robotics' 3D apparatuses to envision their careful plans, read pictures with simulated intellect that perceives functional highlights, and play out a progressively steady and exact spinal activity.
  - MicroSure's robots assist specialists in beating their human physical impediments.

    The organization's movement stabilizer framework allegedly improves execution and exactness during surgeries. As of now, eight of MicroSure's small-scale careful tasks are affirmed for the lymphatic framework methodology.
  - Auris Health builds up an assortment of automatons intended to enhance endoscopies by utilizing the most recent in small-scale instrumentation, endoscope plan, information science, and artificial intellect. Thus, specialists get a more clear perspective on a person's ailment from both a bodily and an information point of view. The organization is creating AI robots to consider lung disease, proposing to fix it sometime in the not so distant future.

- The Accuray CyberKnife System utilizes automated arms to decisively cure carcinogenic tumors everywhere throughout the body. Utilizing the robots, the specialists can treat just influenced territories instead of the entire body.

  The Accuray CyberKnife robot utilizes six-dimensional movement detecting innovation to forcefully track and assault harmful tumors while sparing solid tissue.
- The mechanical autonomy division at Carnegie Mellon University created Heartlander, a smaller-than-expected portable robot intended to encourage the treatment on the heart. Under a doctor's supervision, the little man-made automaton makes its passage to the chest via a little entry point, explores specific areas of the heart without anyone else, holds fast to the outside of the heart, and manages the treatment.
- Natural's da Vinci stages have spearheaded the automated medical procedure industry. Being the principal mechanical medical procedure partner endorsed by the Food and Drug Administration more than 18 years back, the careful machines highlight cameras, automated arms, and careful apparatuses to assistant in insignificantly obtrusive methods. The da Vinci stage is continually learning and giving the investigation to specialists to improve future medical procedures. Up until now, da Vinci has aided more than 5,000,000 tasks.

From the last years, fake wisdom or the so-called artificial intellect in medication has gradually picked up its prevalence. Clinical simulated intellect frameworks, and the innovations for building up these frameworks, have been demonstrated in a tremendous assortment of standard clinical applications.

Besides, as huge scrambled well-being data opens up online for clinical scientists, advanced PC procedures will be expected to utilize the information in a significant manner. Without a doubt soon, numerous clinical applications should incorporate savvy programming parts just to stay serious.

Several areas concerning to the research field in AI are committed to create frameworks that use a blend of the kinds of frameworks introduced above (for example, neural systems and fluffy rationale, master frameworks with coordinated insightful DSS, and so on.). Later on, these mixes of the different AI advancements will keep on being investigated and might be joined to shape a coordinated framework.

6.5 CURRENT RESEARCHES THAT CONTRIBUTE TO THE ADVANCEMENT OF AI

Though a lot has been researched in the field of "AI in Healthcare," still some stones exist that need to be turned using the advancements in the technology. Some of the major domains where the industries and the health professionals are focusing on the current researches are mostly related to life-threatening diseases.

Important disease zones that utilize AI devices include tumor growth, neurology, and cardiology. We survey the AI applications in stroke at that point in more detail, in the main three regions of early detection and conclusion, treatment, as the expectation of result and forecast assessment. Recent trends have shown that an increase in AI researches in medicine-prominent areas under research is briefed beneath.

- Industry
  The ongoing thought pattern of enormous-based well-being organizations converging with other well-being organizations takes more noteworthy accessibility of information about well-being into account [25]. Greater well-being information may take into consideration more execution of AI algorithms [26]. A huge piece of the business focal point of usage of AI in the social insurance division is in the clinical choice help systems [27]. As the quantity of information builds, AI choice emotionally supportive networks become progressively effective. Various organizations are investigating the conceivable outcomes of the joining of enormous information in the social insurance industry [28].
- Imaging
  Recent advancements have prescribed the use of AI to delineate and survey the consequence of a maxillo-facial clinical technique or the assessment of inborn gap treatment regarding the facial charm of age appearance [29]. A paper published in the journal *Annals of Oncology* in 2018 discussed that skin diseases could be perceived even more clearly by a man-made cognizance system (which used a major learning convolutional neural framework) than by dermatologists.
  Overall, 86.6% of skin diseases from the photographs were identified unanimously by the human dermatologists and appeared differently from 95% for the CNN machine [30].
- Radiology
  The ability to translate imaging results with radiology may help clinicians perceive a brief change in an image that a clinician may unintentionally skip.
  In each of those patients, a prevailing typical F1 metric (an accurate estimate subject to exactness and audit) [31] was used in an analysis at Stanford, and the radiologists contributed to that result summary. A few associations (QUIBIM, icometrix, and Robovision) offer AI stages for moving pictures as well [32]. Additionally, their vendor unprejudiced systems such as UMC Utrecht's IMAGR AI. These phases are adaptable through a significant making sense of how to recognize a wide extent of express illnesses and dissipate. During its annual assembly, the radiology gathering Radiological Society of North America has presented on AI in imaging.
  Specific specialists view the advancement of AI development in radiology as a threat because its advancement can achieve improvements in some quantifiable estimates in specific cases, in contrast to authorities [33].
- Formation of New Drugs
  DSP-1181, an OCD medicine molecule (over the top earnest issue) treatment, was developed by modernized thinking via joint projects of Sumitomo Dainippon Pharma (Japanese pharmaceutical firm) and Exscientia (British startup).

Pharmaceutical associations generally experience around five years, while the drug headway took a single year on the same exercises. DSP-1181 was acknowledged for a human starter [34].

- Medication Interactions

  Redesigns in ordinary language arrangement incited the improvement of computations to recognize calm prescription joint efforts in clinical writing. Medication sedate affiliations speak to a risk to those taking various remedies at the same time and the risk rises with the number of drugs taken [35]. To answer the problem of following all known or suspected medicine quiet participation, AI estimations have been made to isolate information on interfacing drugs and their latent capacity impacts from the clinical composition. In 2013, Tries were converged in the DDIExtraction Challenge, in which a gathering of Carlos III University authorities assembled a corpus of composing on a sedate relationship to outline a state-endorsed test for these kinds of calculations. Contenders were taken a stab at their potential to choose among the substance, which meds were seemed to associate, and what the features of their coordinated efforts were. Specialists continue using this corpus to standardize the estimation of the practicality of their calculations [36].

  Various figurings perceive sedate steady participation from structures in customer-made substances, especially electronic prosperity records or possibly ominous event reports. Associations, for instance, the FDA Adverse Event Reporting System (FAERS) and the World Health Organization's VigiBase, grant authorities to submit reports of possible negative reactions to prescriptions. Significant learning counts have been made to parse these reports and recognize structures that propose sedating quiet associations [37].

Simulated intelligence keeps on extending in its capacities, and as it can decipher radiology, it might have the option to determine more individuals to require fewer specialists as there is a lack in huge numbers of these nations [38]. The aim of AI is to improve treatment and in the long run more noteworthy worldwide well-being. Utilizing AI in creating countries that don't have the assets will reduce the requirement for re-appropriating and may utilize AI to improve quiet care. Natural language preparation and AI are used to guide malignant growth medications in spots, for example, India, China, and Thailand. To use NLP to mine through patient records and give treatment, a definitive choice made by the AI application concurred with master choices 90% of the time.

6.6 KEY ISSUES & CHALLENGES AHEAD IN AI

Despite its great potentials, AI presents a whole new set of challenges for the healthcare industry, which includes the following:

- Threats to Data Privacy and Security

  The AI system heavily relies on massive medical data such as EHRs, insurance claim records, sexual preference, genetic information, dietary habits, etc. which help the healthcare provider to improve their service quality and to manage their operations. However, a large data pool of digital data can increase its vulnerability, making them an attractive target, and can threaten the security and privacy of patient data. A lack of appropriate security measures may lead to multiple data breaches, leaving patients suffering from economic and emotional threats. In response to these considerable privacy and security threats of medical data, we need to implement new AI laws, data policy, and regulations.

- Ethical Concern

  Along with the numerous benefits of AI in healthcare, it also raises some major ethical issues such as the potential risk of bias and discrimination into the diagnostic process through the use of databases and algorithms that need to be addressed to ensure the success of these technologies. This might be overcome via data policy and a set of compulsory ethical standards.

  AI-equipped robots are completely logical and unable to feel any sympathy towards the patients and devoid of the sense of moral dignity, compassion, or conservatism. AI should, therefore, function according to a set of values that are consistent with those of humans. Unlike them, physicians can break certain rules to do their utmost to save someone's life. However, a very little research related to the ethical concern of the use of simulated intellect in the medical field and healthcare exists to date.

- Medical Error Risk

  Decision support tools or AI-enabled processes could be wrong, and misleading algorithms are hard to identify. This might led to injury or healthcare problems to thousands of patients across the healthcare system, and since these AI-enabled systems perform complex mathematical transformations to the input data, computer system errors may require extra caution in detection and interpretation. Although AI-enabled processes can identify potential ailments rapidly compared to human physicians, these processes result in poor decision-making. Thus, they cannot yet completely take over for human doctors as there is no room for trial and error when it comes to patient health.

- Reduction in Doctor-Patient Interaction

  Physicians, nurses, and other clinicians deeply care about their relationship with patients and prefer facetime over any other facets of the work. The introduction of AI may interfere with the patient-provider relationship by limiting the interaction between the patient and the clinician, which is important to gain their trust, reassure them, or express empathy. However, this scenario may change in the upcoming days when artificial intellect will be capable of conducting a medical or high-level conversation.

- Threat to Doctors and Healthcare Worker's Jobs

  Humans are subjected to fatigue and clumsiness as a result of confined spaces for long periods. The ability of machines to never get tired, bored, and possession of unlimited stamina seems more desirable to improve healthcare service and increase the threat of robotic dominance in the health domain. Consequently, it has an absolute prospective to replace some doctors such as radiologists, anesthesia, and other healthcare workers in the upcoming future.

6.7 CONCLUSION

The AI-enabled healthcare system will play a critical role in assisting the healthcare service provider to deliver services more effectively in the succeeding years. The complex algorithms need to be programmed and tested before it can assist clinical physicians in the diagnosis of diseases and can recommend any suggestion in treatment through the healthcare data. Accurately programmed AI-equipped systems or instruments should be able to exert a predefined force along the desired direction to obtain the promising results. In the surgical fields, these AI-enabled systems due to their geometrical accuracy and ability to move the instrument in a defined trajectory can completely take over the scope of human error.

Through the rapid development of AI-enabled technology, it can bring a significant improvement in the healthcare system in the following three ways. First, it can bring about an improvement in the productivity and quality of healthcare providers. Second, it can boost patient involvement in their treatment and increases access to care for patients. Third, it reduces the cost of treatments.

While AI technology alone may contribute considerably, the greater potential lies in the synergies created by using them together during the entire journey of a patient from the diagnosis of disease, to treatment to further maintenance of the health of a patient. Also, AI has demonstrated its potential in reading numerous forms of image info incorporating retina scans, ultrasound, and radiographs, and plays a significant role as an informative assistant for understanding the meaning of patterns from data collection and thus can save a lot of time. However, the lack of a large clinical dataset is a significant challenge in healthcare in the training of AI models to perform as required.

Speech and text recognition are now used in activities such as recording of clinical data and to communicate with patients. These free up the time of primary care physicians, which can be utilized to increase productivity and efficiency.

As a conclusion, it can be stated that integrating AI-based technology can be applied for a wide range of purposes, including clinical practices in healthcare. Despite some challenges and issues such as privacy, ethical concern, the threat to healthcare worker's jobs, AI has enormous benefits and compelling evidence to assist the clinician to deliver better healthcare in every aspect of the medical field. With time, AI might be able to overcome the challenge of possessing human abilities such as compassion and motivation, and maybe only the healthcare provider will lose jobs who fails to cope with the artificial intellect. Thus, the application of AI increases convenience and efficiency, and reduces cost and errors.

REFERENCES

1. Poole D, Mackworth A, Goebel R (1998). *Computational Intelligence: A Logical Approach.* New York, NY: Oxford University Press. ISBN 978-0-19-510270-3.
2. Russell SJ, Norvig P (2009). *Artificial Intelligence: A Modern Approach* (3rd ed.). Upper Saddle River, NJ: Prentice Hall. ISBN 978-0-13-604259-4.
3. McCorduck P (2004). *Machines Who Think* (2nd ed.). Natick, MA: A. K. Peters, Ltd.. ISBN 1-56881-205-1.

4. Crevier D (1993). *AI: The Tumultuous Search for Artificial Intelligence.* New York, NY: Basic Books, ISBN 0-465-02997-3.
5. Davenport T, Kalakota R (2019). The potential for artificial intelligence in healthcare. *Future Healthcare J* 6(2): 94.
6. Algorithms need managers, too. *Harvard Business Review.* 1 January 2016. Retrieved 2018-10-8. https://hbr.org/2016/01/algorithms-need-managers-too
7. Deloitte Insights (2018). State of AI in the enterprise. *Deloitte.* https://www2.deloitte.com/content/dam/insights/us/articles/4780_State-of-AI-in-the-enterprise/AICognitiveSurvey2018_Infographic.pdf.
8. Lee SI, Celik S, Logsdon BA, Lundberg SM, Martins TJ, Oehler VG, Estey EH, Miller CP, Chien S, Dai J, Saxena A (2018). A machine learning approach to integrate big data for precision medicine in acute myeloid leukemia. *Nat Commun* 9: 42.
9. Fakoor R, Ladhak F, Nazi A, Huber M (2013). Using deep learning to enhance cancer diagnosis and classification. A conference presentation. *The 30th International Conference on Machine Learning*, Atlanta.
10. Sordo M (2002). Introduction to neural networks in healthcare. *OpenClinical.* https://www.openclinical.org/docs/int/neuralnetworks011.pdf.
11. Vial A, Stirling D, Field M, Ritz C, Carolan M, Holloway L, Miller AA (2018). The role of deep learning and radiomic feature extraction in cancer-specific predictive modelling: a review. *Transl Cancer Res* 7: 803–816.
12. Hussain A, Malik A, Halim MU, Ali AM (2014). The use of robotics in surgery: a review. *Int J Clin Pract* 68: 1376–1382.
13. Davenport TH, Glaser J (2002). Just-in-time delivery comes to knowledge management. *Harvard Business Review.* https://hbr.org/2002/07/just-in-time-delivery-comes-to-knowledge-management.
14. Dilsizian SE, Siegel EL (2014). Artificial intelligence in medicine and cardiac imaging: harnessing big data and advanced computing to provide personalized medical diagnosis and treatment. *Curr Cardiol Rep* 16: 441.
15. Jiang F, Jiang Y, Zhi H, Dong Y, Li H, Ma S, Wang Y, Dong Q, Shen H, Wang Y (1 December 2017). Artificial intelligence in healthcare: past, present and future. *Stroke Vasc Neurol* 2(4): 230–243.
16. COMP3330: History assignment. http://www.angelfire.com/ks2/kaz/ai_medicine/timeline.html#conc.
17. Lindsay RK, Buchanan BG, Feigenbaum EA, Lederberg J (1993). DENDRAL: a case study of the first expert system for scientific hypothesis formation. *Artificial Intelligence* 61(2): 209–261. doi: 10.1016/0004-3702(93)90068-m.
18. Clancey WJ, Shortliffe EH (1984). *Readings in Medical Artificial Intelligence: The First Decade.* Boston, MA: Addison-Wesley Longman Publishing Co., Inc.
19. Bruce G, Buchanan BG, Shortliffe ED (1984). *Rule-Based Expert Systems: The MYCIN Experiments of the Stanford Heuristic Programming Project.* Reading, MA: Addison Wesley.
20. Duda RO, Shortliffe EH (April 1983). Expert systems research. *Science* 220(4594): 261–268. doi: 10.1126/science.6340198. PMID 6340198.
21. Miller RA (1994). Medical diagnostic decision support systems – past, present, and future: a threaded bibliography and brief commentary. *J Am Med Inf Assoc* 1(1): 8–27. doi: 10.1136/jamia.1994.95236141. PMC 116181. PMID 7719792.
22. Adlassnig KP (1980). A fuzzy logical model of computer-assisted medical diagnosis. *Methods Inf Med* 19: 14.
23. Baxt WG (December 1991). Use of an artificial neural network for the diagnosis of myocardial infarction. *Ann Intern Med* 115(11): 843–848. doi: 10.7326/0003-4819-115-11-843. PMID 1952470.

24. 32 Examples of AI in healthcare that will make you feel better about the future. https://builtin.com/artificial-intelligence/artificial-intelligence-healthcare.
25. La Monica PR (2018). What merger mania means for health care. *CNNMoney*. Retrieved 2018-4-11.
26. Why you're the reason for those health care mergers. *Fortune*. Retrieved 2018-4-10. https://fortune.com/2018/03/19/cvs-aetna-healthcare-mergers-big-data/
27. Horvitz EJ, Breese JS, Henrion M (July 1988). Decision theory in expert systems and artificial intelligence. *Int J Approx Reason* 2(3): 247–302. doi: 10.1016/0888-613x(88)90120-x. ISSN 0888-613X.
28. Arnold D, Wilson T (June 2017). What doctor? Why AI and robotics will define New Health (PDF). *PwC*. Retrieved 2018-10-8. https://www.pwc.com/gx/en/news-room/docs/what-doctor-why-ai-and-robotics-will-define-new-health.pdf
29. Patcas R, Timofte R, Volokitin A, Agustsson E, Eliades T, Eichenberger M, Bornstein MM (August 2019). Facial attractiveness of cleft patients: a direct comparison between artificial-intelligence-based scoring and conventional rater groups. *Eur J Orthod* 41(4): 428–433. doi: 10.1093/ejo/cjz007. PMID 30788496.
30. Computer learns to detect skin cancer more accurately than doctors. *The Guardian*. 29 May 2018. https://www.theguardian.com/society/2018/may/29/skin-cancer-computer-learns-to-detect-skin-cancer-more-accurately-than-a-doctor
31. Rajpurkar P, Irvin J, Zhu K, Yang B, Mehta H, Duan T, Ding D, Bagul A, Langlotz C, Shpanskaya K, Lungren MP (14 November 2017). CheXNet: radiologist-level pneumonia detection on chest X-rays with deep learning. ArXiv: 1711.05225 [cs.CV].
32. Chockley K, Emanuel E (December 2016). The end of radiology? Three threats to the future practice of radiology. *J Am Coll Radiol* 13(12 Pt A): 1415–1420. doi: 10.1016/j.jacr.2016.07.010. PMID 27652572.
33. Jha S, Topol EJ (December 2016). Adapting to artificial intelligence: radiologists and pathologists as information specialists. *JAMA* 316(22): 2353–2354. doi: 10.1001/jama.2016.17438. PMID 27898975.
34. Artificial intelligence-created medicine to be used on humans for first time. *BBC News*. 30 January 2020. https://www.bbc.com/news/technology-51315462
35. García Morillo JS (2013). Optimización Del tratamiento de enfermos pluripatológicos en atención primaria UCAMI HHUU Virgen del Rocio. Sevilla. Spain. Available for members of SEMI at: ponencias de la II Reunión de Paciente Pluripatológico y Edad Avanzada Archived 2013-4-14 at Archive.today.
36. Christopoulou F, Tran TT, Sahu SK, Miwa M, Ananiadou S (2020). Adverse drug events and medication relation extraction in electronic health records with ensemble deep learning methods. *J Am Med Inform Assoc* 27(1): 39–46.
37. Xu B, Shi X, Yin Y, Zhao Z, Zheng W, Lin H, Yang Z, Wang J, Xia F (July 2019). Incorporating user generated content for drug drug interaction extraction based on full attention mechanism. *IEEE Trans Nanobioscience* 18(3): 360–367.
38. Chen AF, Zoga AC, Vaccaro AR (1 November 2017). Point/counterpoint: artificial intelligence in healthcare. *Healthcare Transform* 2(2): 84–92. doi: 10.1089/heat.2017.29042.pcp.

7 Artificial Intelligence and Personalized Medicines

A Joint Narrative on Advancement in Medical Healthcare

Shyam Bass and Sheetu Wadhwa
Lovely Professional University

CONTENTS

7.1 INTRODUCTION

Precision medicine is a progressing healthcare advancement that focuses on tailoring medical products, treatments, and medical practices in accordance with patients who are having variability in genetics, surroundings, living style, and other factors that are providing the more righteous treatment at the right time to the right patient. Artificial intelligence (AI) refers to the understanding skills by computers or high-technology machines that imitate the known functions that are associated with the humans or human mind, such as understanding, explanation, composition, and solving of problems (Hulsen et al. 2019).

There is an evolution in evaluating power, conceptual understanding, and increasing piles of data, over the last few decades; hence, scientists observed an extensive approach of AI in almost every field of society—majorly in medicine and healthcare

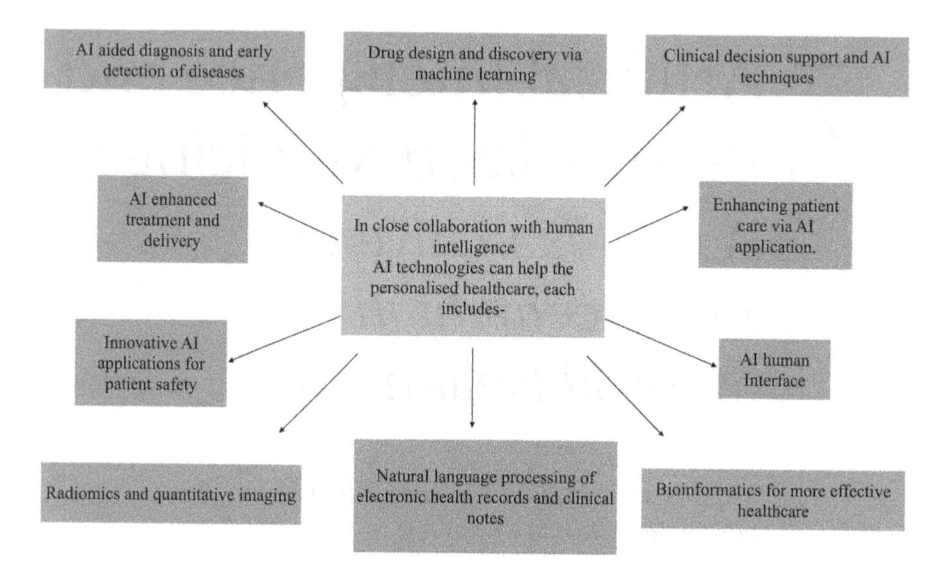

FIGURE 7.1 Artificial intelligence in personalized healthcare.

sectors. AI can support in the modernized healthcare with the concern of precision medicine in three major classes: (a) advanced preventives for diseases, (b) individualized diagnosis, and (c) unique treatment.

This chapter is focused on the development of precision medicines with the application of AI to the modern healthcare (Hulsen et al. 2019; Piccart-Gebhart et al. 2005). Moreover, these AI technologies in close collaboration with human intelligence aid in the personalized healthcare worldwide in a more effective manner (as shown in Figure 7.1).

7.2 NEED FOR PERSONALIZED MEDICINES

Personalized medicine is a tailor-made approach towards the medical treatment as per the unique and understandable characteristics of an individual. The scientific development depends upon the approach of our understanding that how distinctive molecular profile and genetic profile of a person make them suffered from few diseases. The AI technique is providing the prediction ability, which is the safest and most effective medical treatment for each patient and shown in Figure 7.2.

The concept of personalized medicine equipped with a system is more precise. Hence, a medical practitioner can select a particular tailor-made treatment approach based upon the molecular-level understanding of patients—this approach lowers the chances of side effects, which may be harmful, and not only gives a more successful result but can also decrease the use of the "trial-and-error"-based treatment approach. Personalized medicine has the probability to change the way of thought process of identifying and managing health problems.

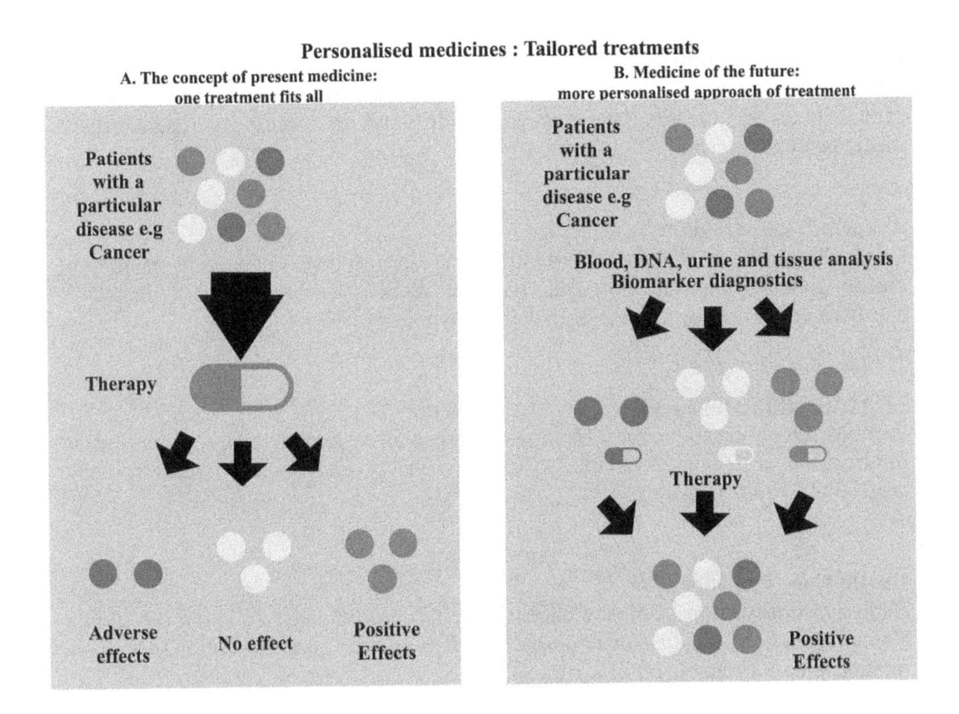

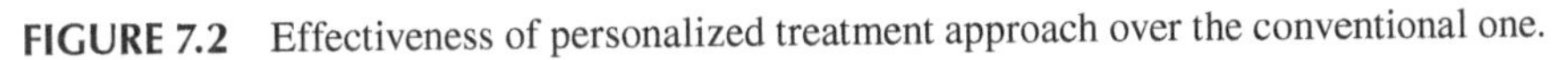
FIGURE 7.2 Effectiveness of personalized treatment approach over the conventional one.

This concept is creating an extraordinary impact in the area of better patient care and improved clinical research owing to improved technologies and enhancement in knowledge (Bertalan 2017; Mathur and Sutton 2017).

7.2.1 Contributors to Personalized Medicines

The production and realistic approach of personalized medicine relies on the resources and contributions of a broad community of different sectors, all working together toward a shared goal of forming breakthroughs in science and technology to improve the patient care (Mathur and Sutton 2017).

These contributors are in the form of patients/consumers, academic researchers, IT specialists, healthcare providers, data interpreters, and biopharmaceutical-based companies. Their contributions are depicted in Figure 7.3 and discussed as follows:

(a) Patients and consumers who participate in genetic-level testing and clinical trials, and provide a support in the management of treatment; (b) academic researchers who conducted a basic and applied clinical research to understand the disease loci at the molecular level and as a result, developed an appropriate drug formulation and enabled the better diagnostic approaches; (c) IT section who creates devices, tools, and resources to collect and store patients' health information, which further helps in taking clinical decisions while maintaining privacy of the data; (d) healthcare providers who gather a genetic makeup information of an individual and utilize this

Patients and Consumers:
The patients participates in genetic testing and clinical trials and support the health care workers in managing the treatment contributes in success of a personalised medicine development.

Academic Researchers:
Conducting basic fundamentals and clinical research to reveal new understanding of human genetics and the molecular basis of disease, enabling greater diagnostic approach and more appropriate drug development.

IT/Artificial Intelligence:
Different electronic devices, tools and resources to collect and store patient health information, which are available to improve safety and clinical decisions while protecting patient privacy.

Health Care Providers:
Gathering information from the patient's genetic profile and by making use of new technologies to personalise this approach to disease detection, prevention, diagnosis, treatment, and management of treatment.

Diagnostic /Data interpretation :
Variable developing tools and experiments for analyzing and interpreting genetic information, improves the understanding of disease at the molecular level. This is how a patient shows more likely response towards the drug.

Biopharmaceutical Companies:
Developing targeted treatment and performing novel research based upon knowledge of genetic variation and its effects on the safety and effectiveness of the drug on the candidate.

FIGURE 7.3 Contributions of different sectors in the development of personalized medicines.

data in detection, diagnosis, prevention, management, and personalized treatment; and (e) data interpreter who analyzes the data.

7.3 APPLICATION OF AI IN HEALTHCARE FOR DEVELOPMENT OF PRECISION MEDICINES

From the last two decades, the concept of individualized medicines has become popular, but in the last few years, the genetic-level patient data and database of electronic health records have been emerged and proven helpful to doctors towards the development of personalized medicine as per the diagnosis of an individual patient. Moreover, Dr. Bertalan Meskó who is the director at Medical Futurist Institute recommended the AI in healthcare and states that "there is no precision medicine without AI," and without AI techniques, the analysis of patient database will remain untapped (Bertalan 2017).

Noticing the increasing needs in the healthcare system and indeed the development of individualized treatments, the genetic data is one of the key points of discussion in order to study and store the data (Douali and Jaulent 2012). And for this requirement, there is only one meal course that is AI, which is going to provide a complete nutrition to the hunger and solve the problem.

The multidata sources of AI like genetic data, electronic health record data, environmental data, and lifestyle data (as shown in Figure 7.4) are the basis of AI and are putting steps forward for the development of personalized treatments for the diseases such as from the depression to the cancer, although there are many challenges which researchers have to face but this advanced approach towards healthcare can prove to be a boon for society (Bertalan 2017).

The following points elaborate the need of AI in the development of precision medicines:

- AI made it possible to quickly sequence the whole human genome using the next-generation sequencing.
- With AI, it is now even more possible by utilizing technology to develop a personal genome sequence at very reasonable price in the future, which

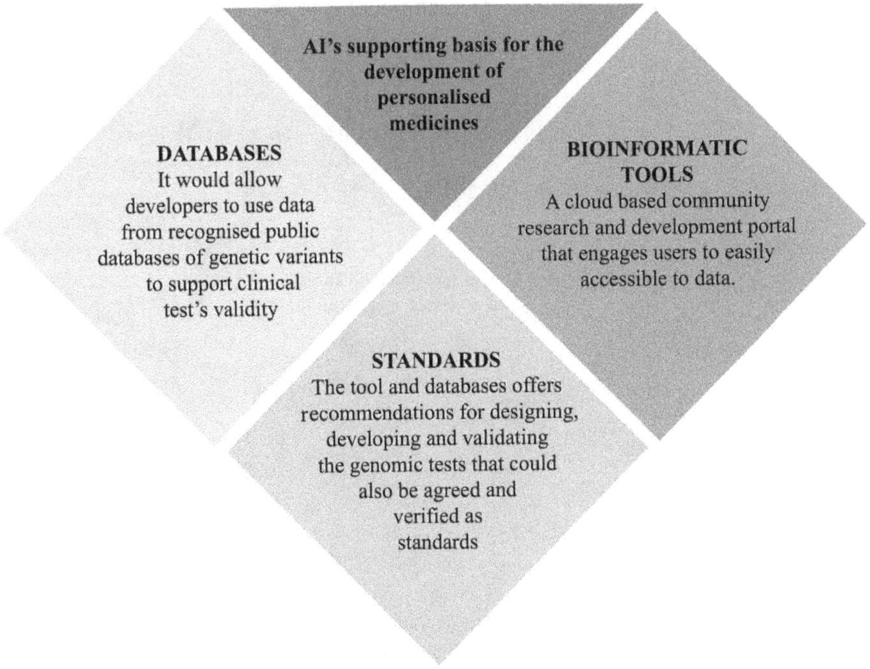

FIGURE 7.4 Artificial intelligence support in personalized medicines.

made the analysis process easier with genome sequence, which was earlier difficult and very time-consuming too.

- The study of genome sequencing with the collaboration of AI allows a better and easy diagnostics for the development of unique medicines in the future.
- With AI, the work which has been done earlier by the scientists has been stored in the large databases which are helping today's scientists to study and analyze the large database in a more efficient manner for the development of personalized medicine Ginsburg and McCarthy 2001; Mesko 2017).

7.4 IN INTENSIVE CARE UNIT (ICU)

7.4.1 In Intensive Care Unit (ICU)—To Predict the Fluid Requirement

Few activities that are performed in the ICU are examination of disease, scanning, and handling critical conditions and treatment. To predict the fluid requirement in the ICU with the help of experimental data to obtain patient- and clinical scenario-related recommendations, as discussed in Figure 7.5, the software applications are revised and reviewed that form a patient index, which is similar to the standard patient index in terms of length of life lived, sex, tradition, admitting diagnosis, severity on admission, and co-occurring diseases.

Based on the model representation, the physiological variability studies that are directly related to the outcomes are identified. And the concept is to get the values of these predictor variables from the index patient, and then, a predicted range of fluid requirement is obtained from the joint conditional probabilities (Celi 2008).

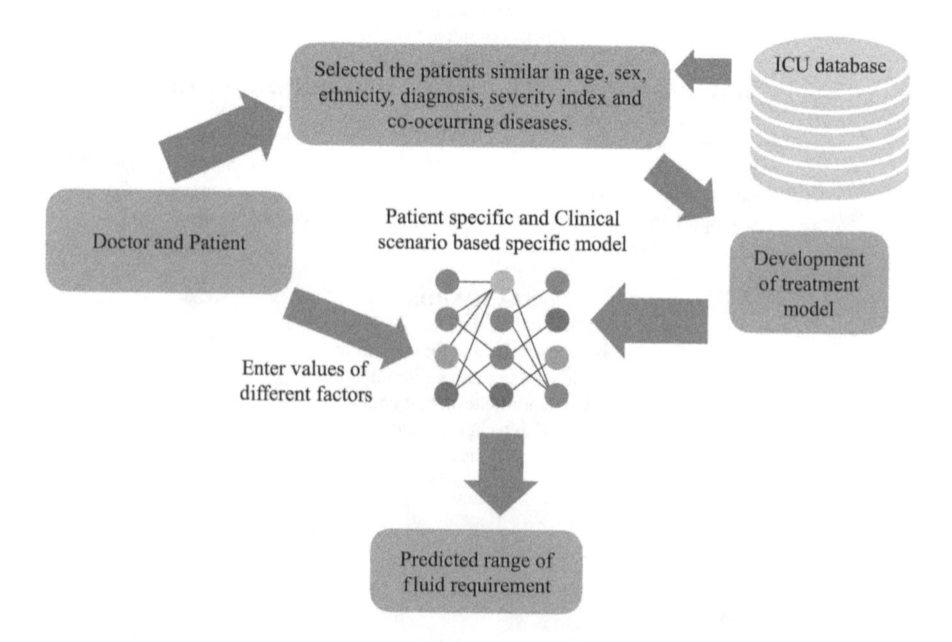

FIGURE 7.5 Role of artificial intelligence in predicting fluid requirements in ICU.

For example, the Multi-parameter Intelligent Monitoring for Intensive Care II (also known as MIMIC II) is a patient database for patients who are admitted to Beth Israel Deaconess Medical Centre, Boston, and this database identified the specific patients who were on vasopressors for more than 6 hours during the first 24-hour time period (Celi 2008). Based on demographic and physiological factors, there is a variation in fluid requirement or difference in the intravascular fluid volume for the last 24 hours. Hence, planning is done as per the availability of the information in ICU by the AI system, and then, the treatment is predicted by the observational study (Celi 2008).

To replace clinical expertise with intelligent software is so improbable, as the observational and concluding part is still under the human intelligence. There are other important applications of the AI tools in data observation, and these tools are available: (a) to utilize the clinical knowledge to analyze the particular or complicated problems, (b) to accelerate the diagnosis and treatment approach adopted by the doctors, and (c) to monitor the results obtained from the clinical outcomes, especially in the ICU.

7.4.2 To Solve Issues of Personalized Medicines

Personalized medicines are an extension of medical science field to provide the customized healthcare services to patients. In personalized medicines, the main components are to predict the possibility that an individual may have a chance of developing a disease, then to achieve the accurate diagnosis and the best possible treatment of that particular disease. Various linear and nonlinear models such as Naïve Bayesian (NB), artificial neural network (ANN), and support vector machines (SVMs) are also adopted to produce the accurate results from the variable input data (Awwalu 2015).

7.4.3 Revolutionizing Cloud of AI and Healthcare

With the drastic transformation into the digitalized healthcare system, several advanced technologies have emerged, which are available for clinicians, healthcare professionals as well as for patients. The developed technologies based on AI such as genome sequencing, biotechnology-based product, and wearable sensors are gradually leading to an advancement in healthcare field. Moreover, taking care of patients and their treatment as per their unique characteristics, and then creating a large amount of data requiring advanced analytics and the foundation of precision medicine are other advantages (Bertalan 2017).

With AI, there is a gradual shift towards the prevention, personalization, and precision among patients instead of taking the old treatments applied on populations with the similar physical characteristics. Now need of the hour is to bring this opportunity into daily practice, as shown in Figure 7.6 (Awwalu 2015; Jason 2017; Bresnick 2018).

7.5 CONCLUSION

AI inspired machine learning technologies to develop the most advanced system for innovating the today's need on medicine, i.e., precision medicine. Biological and medical sciences of genomics, image processing, and discovery of drugs rapidly

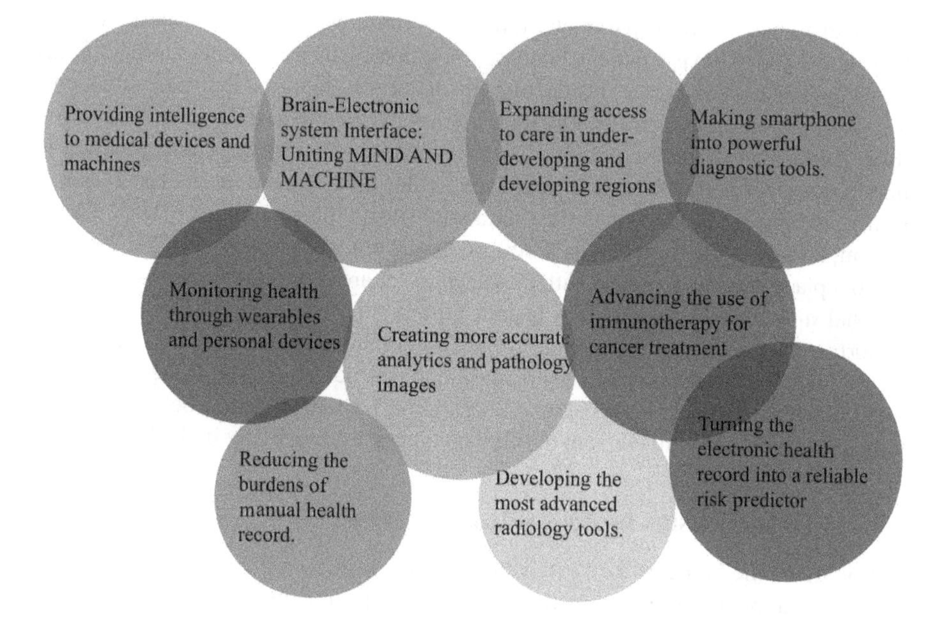

FIGURE 7.6 Applications of artificial intelligence in healthcare.

accepted the AI technology in their system. From traditional approaches to robotics, the introduction of several technologies has made the medical healthcare system easier. The high-speed and more accurate machines and their networks have changed the approach towards analyzing the medical bioinformatics at the unmatched speed. As a result, the process of decision-making in the development of precision medicine will still to be more evolved but AI has given a considerable shift in the thought process towards the health sector.

REFERENCES

Awwalu, J., Garba, A., Ghazvini, A., Atuah, R., 2015. Artificial intelligence in personalized medicine application of AI algorithms in solving personalized medicine problems. *International Journal of Computer Theory and Engineering* 7(6), 439–443.

Bresnick, J., 2018. Top 12 ways artificial intelligence will impact healthcare. Retrieved from https://healthitanalytics.com/news/top-12-ways-artificial-intelligence-will-impact-healthcare.

Celi, L.A., Christian, L.H., Alterovitz, G., Szolovits, P., 2008. An artificial intelligence tool to predict fluid requirement in the intensive care unit: a proof-of-concept study. *Critical Care* 12, 1–7.

Douali, N., Jaulent, M.C., 2012. Genomic and personalized medicine decision support system, in *Proceedings of the IEEE International Conference on Complex Systems*, 1–4. IEEE, Agadir, Morocco.

Ginsburg, G.S., McCarthy, J.J., 2001. Personalized medicine: Revolutionizing drug discovery and patient care. *Trends in Biotechnology* 19(12), 491–496.

Hulsen, T., Jamuar, S., Moody, R., Karnes, H., Varga, O., Hedensted, S., Spreafico A., McKinney, F., 2019. From big data to precision medicine. *Frontiers in Medicine* 6, 34. doi: 10.3389/fmed.2019.00034.

JASON, 2017. Artificial intelligence for health and health care. Retrieved from https://fas.org/irp/agency/dod/jason/ai-health.pdf.

Mathur, S., Sutton, J., 2017. Personalized medicine could transform healthcare. *Biomedical Reports* 7(1), 3–5.

Mesko, B., 2017. The role of artificial intelligence in precision medicine. *Expert Review of Precision Medicine and Drug Development* 2(5), 239–241.

Piccart-Gebhart, M.J., Procter, M., Leyland-Jones, B., Goldhirsch, A., Untch, M., Smith, I., Gianni, L., Baselga, J., Bell, R., Jackisch, C., Cameron, D., 2005. Trastuzumab after adjuvant chemotherapy in HER2-positive breast cancer. *The New England Journal of Medicine* 353(16), 1659–1672.

8 Nanotechnology and Artificial Intelligence for Precision Medicine in Oncology

Tajunisa M., L. Sadath, and Reshmi S. Nair
Amity University Dubai

CONTENTS

8.1 INTRODUCTION

8.1.1 Fundamentals of Nanotechnology

Nanotechnology is the science dealing with the modification of materials at an atomic or a supramolecular level. Considering the last few decades, the fundamentals of nanotechnology have emerged as an autonomous branch of science with explicit research opportunities to the world of science and society. The interdisciplinary approach, inquisitiveness of research, and greater public awareness have established nanotechnology as the most desirable field. Recent studies in nanotechnology reveal the evident prospects of the achievements, which are only a tip of an iceberg to the scientific fraternity and many more to come as stated by Dr. Feymann: "There is plenty of room at the bottom." Hence, the scientific world is awaiting this technology as "the technology of the next century" [1].

The novelty and exclusivity of nanotechnology is apprehended by the amazing properties exhibited in the nanoscale objects. When the materials are reduced to a nanolevel with the phenomenon of quantum confinement, the materials tend to invade larger surface area, thereby increasing the surface energy and exhibiting various advanced properties. Nanomaterials have proven the concept that size really matters for any object when it gets converted from a bulk material to the nanoscale with magical properties. The effectiveness of the size is dependent on the properties of a material like optical, electrical, magnetic, melting points, specific heat, biological mechanisms, surface reactivity, and so on [2].

Extensive research in this field of science has led the community of scientists and academicians to accept the potential of nanotechnology for the well-being of mankind and to enhance the standard of life [3,4].

The impact of nanotechnology is getting extended to every aspect of human life due to its product application across various prominent industries like healthcare, biomedical, pharmaceutical, electronics, and energy production. As the manufacturing and processing mechanisms are dealt with engineering concepts, the current challenge is faced by engineers to develop a novel strategy with the nanotechnology for new products in the market. Besides understanding the fundamentals of this wonderful technology, this field demands for more creative ideas for developing technology. Moreover, critical thinking and creativity are considered the key features during the learning process of nanotechnology. Due to these significant characteristics, it is considered as an interdisciplinary platform to discover the opportunities in engineering sciences and information technology with a wider aspect to benefit the society.

8.2 ROLE OF NANOTECHNOLOGY IN MEDICINE AND HEALTHCARE

Nanotechnology plays a significant role in healthcare and translational medicine. Exploiting nanoparticle sizes and nanosized gaps between structures represent other ways of obtaining new properties and physical access inside the tissues and cells. Quantum dots are used for visualization in drug delivery because of their fluorescence and ability to trace very small biological structures. The secondary effects of the new techniques include raising safety concerns such as toxicity that must be addressed before the techniques are used in medical practice. Emerging areas include developing realistic molecular modeling for "soft" matter [5], obtaining non-ensemble-averaged information at the nanoscale, understanding energy supply and conversion to cells (photons and lasers), and regeneration mechanisms.

Cancer is one of the leading causes of death in today's world. According to 2017 Cancer statistics, approximately 9.6 million people died due to different forms of cancer [6]. Nanotechnology has proven to be one of the most efficient methods of drug delivery. Nanoparticles—given their surface area-to-volume ratio, their size, and their shape—allow multiple functionality and easy alteration of their properties. They can be programmed to differentiate between malignant cancer cells and healthy human cells [6]. Nanotechnology allows a rapid testing of the diseases through imaging techniques, thus making sure that these diseases are detected at an earlier stage to allow the use of less-evasive techniques needed to cure them at their root or just cause minimum damage. This section will describe various applications and their unique properties that make nanoparticles in different forms apt for their function.

Exosomes are a type of extracellular vesicles and are useful mediators in the long-distance intercellular communication. A well-modified biocompatible exosome increases the stability and efficacy of the imaging probes with an enhanced regulatory mechanism to overcome the cell membrane [7]. Nanosized exosomes are useful in identifying biomarkers indicating pancreatic cancer. This is based on the theory that a nanoparticle carries several peptide markers, which are found in high concentration at an early stage as well and hence can be detected [8].

Similarly, nanopore sensors and Artificial intelligence (AI) technology are utilized to identify single-virus particles for the point of use and rapid identification. A nanopore is a pore of nanometer size; it is based upon the principle that when a flow of liquid is applied across its membrane, it will measure the ionic current passing through its pore. This sensor is extremely sensitive allowing only viral particles that are specific to the sensor to pass through it. AI is used to identify the type of viral substance that passes through the nanopore. This technology detects what is left out by the naked human eye when analyzing a current waveform difference, therefore allowing a high-precision identification as mentioned earlier [8]. Another nanosensor application used in cancer diagnosis is a method for the early detection of a disease, which uses nanoparticles that form clumps when they attach to proteins or other molecules that indicate the disease being tested for. This magnetic detection nanotechnology technique is intended to be inexpensive and simple to perform, detecting a biomarker as small as 30 molecules per cubic centimeter of blood. The solution turns blue if the nanoparticles are clumped around a protein indicating the disease; the solution is red if the protein is not present. This technique allows the doctors to realize quickly and rightfully whether the treatment is working or not [9].

A nanoflare is a spherical nucleic acid (SNA) with a gold nanoparticle at the center and densely packed oligonucleotide shells with a single DNA "flare." This was the first genetic approach technique that isolates and detects live cancer cells from the complex matrix that is human blood itself. Nanoflares have a rapid cellular uptake [10]. "We've taken perhaps the world's most important molecule, DNA, rearranged it into a spherical shape and modified it to detect specific molecules inside cells," says Dr. Chad Mirkin, a researcher at Northwestern University and one of the developers of this diagnostic tool [11]. Nanoflares are used to detect cancer in the bloodstream; the markers light up because of the highly efficient fluorescent property of gold when they come in contact with the desired intracellular target mRNA [12]. Along with this technique, a gold nanoparticle coated with a monoclonal antibody is employed in the diagnosis of cancer as well. These coated nanoparticles are produced by the process of plasma vapor deposition. The gold nanoparticle coating forms covalent bonds to bind with other specific target antibodies found in the human body, especially those that express themselves much more such as cancer cells and angiogenesis. They work like nanoflares, where the amount of reflected light tells one how much of the sample contains virus particles [13,14]. There are also other antibody-coated nanoparticles that follow the similar technique of binding. Nanofibers coated with antibodies bind to cancer cells, trapping these neoplastic cells to be analyzed further. Nanofibers are very flexible and have a very versatile polymer composition, porosity, morphology, and easy functioning on the surface. They also have a highly porous and interconnected structure. They are especially used to capture individual cancer cells [13]. These nanofibers can be loaded with drugs in their fibrous matrices: to deliver these drugs into the system in order to kill the residual tumor cells, to detect the tumor cells, and to get them embedded into the tumor bed [15]. Carbon nanotubes and gold nanoparticles have been used to detect oral cancer in less than an hour [15]. Carbon nanotubes are one-dimensional as well, which are biocompatible and transportable in biological fluids. Peptides, nucleic acids, and antigens are transported with high order and regulation. Their action isn't dependent upon the cell type and functional group on the surface of the cell [16].

Another form of a nanoparticle used is a nanowire. The nanowires are a one-dimensional system, which have a cross-sectional area in the nanometer range and the length and width in the 1000 nm range. As there is a high surface area-to-volume ratio of a nanomaterial, this allows the material to have quantum electrical, mechanical, and physical properties. Some of the properties include fast detection and high-speed data delivery. These nanowires are widely used for their semiconductor properties in terms of their applications. The smaller the structure of the nanomaterial, the more control and ease of use a researcher has. The nanowire sensor is defined as: "EVs are potentially useful as clinical markers. The composition of the molecules contained in an EV may provide a diagnostic signature for certain diseases," the lead author Takao Yasui explains. "The ongoing challenge for physicians in any field is to find a noninvasive diagnostic tool that allows them to monitor their patients on a regular basis" used for diagnosing bladder and prostate cancer in urine samples [17]. Similar to the concept of high surface area-to-volume ratio of nanomaterials, we have nanorods. Nanorods are also one-dimensional structures. Silver nanorods give a longitudinal surface plasmon resonance in the spectral range of 400–700 nm. Silver nanorods allow for a better detection of separated components of the blood samples like viruses, bacteria, and microscopic components by employing a better surface-enhanced Raman spectroscopy (SERS) signals in a quick manner. The signals are highest when the plasmon resonance of the metal matches with the laser excitation wavelength produced by the source [18,19]. Another application of nanoparticles allowing enhanced spectrophotometry techniques is when iron oxide nanoparticles are used to develop better images from magnetic resonance imaging (MRI) scans [20]. MRI scans generally use a dye of some sort to allow a formation of contrast between the cells. They are widely used for clinical diagnosis without actually invading the human body. The iron oxide nanoparticles help in the detection of cancer cells, cardiovascular disease, and even neurological diseases, all by functioning on the cellular level. Hence, they make very effective contrast agents in MRI scanning [21]. Similarly, magnetic nanoparticles and nuclear magnetic resonance (NMR) spectroscopy can be related: microvesicles are formed when particles attach to magnetic nanoparticles. Microvesicles originate in brain cancer cells. NMR is used to detect these magnetic nanoparticle clusters; therefore, early diagnosis is possible [22].

Quantum dots are used in cancer cell imaging. They have both in vivo and in vitro techniques that can be used. They are proven to be beneficial because of their unique electronic and optical properties. Quantum dots locate cancer tumors and perform diagnostic tests in samples with high precision and sensitivity. Cadmium quantum dots, which were used initially, are more toxic, compared to the recently developed silicon ones. They are proven to target biomolecules using different bypass routes resulting in an enhanced detection, and the integration of thousands of labels into one nanoparticle results in a drastically amplified signal. All these processes can be monitored noninvasively in a real time [23].

8.2.1 Nanodrug Design by AI

Nano-encapsulated drug system has played a significant role in medicine and healthcare. The primary requirement for a precision medicine could be well connected with the profile in regard to the molecular system of a patient. The role of disease-specific biomarkers is inevitable to provide a clarity and an idea for a personalized treatment. The genetic biomarkers include the various forms of polymorphism of DNA present in the human genetic sequence. Based on these variants like single nucleotide polymorphism, restriction fragment length polymorphism, microsatellite etc., we can analyze the genetic sequence. To apprehend the distinct molecular identity of the patient, a disease profile is useful. This design includes genomic, epigenomic, transcriptomic, proteomic, and microbiomic metabolomics data [24]. In population-wide omics, it is significant to access the recognition process of relevant disease biomarkers and their distribution among the broad spectrum of patients analyzed. For instance, RNA-based molecular profiling helps in differentiating populations with a wide classification of healthy community and patient with localized and metastatic tumor confirmed in their body [25]. RNA sequencing is a standard methodology adopted for analyzing the molecular profiling of the samples from patient with tumor. The accuracy in localization of the tumor is confirmed with the aid of RNA sequencing with tumor-educated platelets (TEP) and blood platelets with altered RNA profile. It is essential to recognize new biomarkers efficiently with the help of accurate and rapid data collection tools.

Nanotechnology enhances the precision of sequencing technologies utilized for various data collection profiles. At the same time, nanotechnology improves the speed of data analysis with better sequence modality and diagnosis tools. For example, single-molecule real-time (SMRT) sequencing and nanopore sequencing methods are the third-generation sequencing techniques proven to be more efficient. This mechanism involves a direct analysis of single DNA nucleotide without amplification of the template, thereby reducing the reading errors [27]. SMRT system utilizes the electron beam lithography method to prepare cavities of the size 60–120 nm. This has been developed on a thin aluminum sheet of 100 nm size deposited on a substrate made of silica. In each of the cavity, DNA polymerase enzyme is embedded to obtain the specific observation during optical imaging with fluorescent dye. Thus, the real-time data is obtained with precision by maintaining the genetic property of the nucleotide and overcoming the challenges in sequencing of nucleic acid [28,29]. In the nanopore sequencing method, the application of AI for translation process of the raw signal to nucleotide sequence is another major impact of interdisciplinary approach of nanotechnology. Hence, the accuracy of patient diagnosis with biomarker profiling has enhanced predominantly with the combination of omics data from various biological sources to generate unified profiles.

8.2.2 Artificial Intelligence

AI is a broad branch of computer science dealing with the simulation of human intelligence in machines that are programmed to exhibit the traits associated with humans. The term was coined by John McCarthy during a conference in 1956. The goal of AI is to create systems that can function both intelligently and independently. In the context of a human, learning and problem-solving by mimicking their actions are categorized as intelligence demonstrated by machines. These devices perceive their surrounding environment and take appropriate measures to increase the chances of success in achieving goals. A more elaborate definition characterizes AI as "a system's ability to correctly interpret external data, to learn from such data, and to use those learnings to achieve specific goals and tasks through flexible adaptation" [30]. This study of intelligence can be classified as a multidisciplinary subject as it can be viewed from many perspectives such as linguistics, philosophy, psychology, mathematics, and medicine as each of these has an influence when attempted to model human intelligence on a machine.

There are two ways in which AI works: symbolic-based and data-based. Machine learning (ML) or data-based AI requires ample information or high-dimensional data that is fed to the machine in order to determine the patterns to make classifications or predictions and solve problems. This technology is rapidly adopted in many fields essentially for precision, improved performance, time efficiency as well as reduced cost.

Today, this technology is incorporated into our lives in the forms of personal assistants (Siri, Alexa etc.), aviation, and computer gaming. It has also been integrated into medicine to improve the patient outcomes in order to provide a better healthcare by enhancing the process and reaching great accuracies in radiological images and so on.

8.2.2.1 AI in Medicine

The two most important factors that are important to physicians are knowledge and experience as the number of patients you treat increases, the more you know and provide better care. These aspects are gained throughout one's career, which may take time and effort through continuous education in their specific area of interest. Better knowledge-based decisions can be established with adequate data derived from evidence-based medicine (textbooks, manuscripts etc.) and experience gained through the real-life outcomes by the treatment of patients, which includes patient files, laboratory results, and radiological images. The acquisition of huge amounts of data poses a primary limitation of a human mind and the process of learning and experience obtained along the years. However, this concept is the underlying principle to understanding AI and its inference in medicine [31]. Harnessing vast amounts of data, transforming them into experience in lieu of wasting years acquiring it, and developing algorithms in significantly short time with improved accuracy are the cornerstones of AI [32]. In medicine, this technology is aimed to improve patient outcomes via early detections and diagnosis by reducing medical errors and costs in addition to a decrease in morbidity and mortality.

The healthcare industry today faces numerous challenges: the growth and aging of population, growing complexity, and escalating costs. These systems are facing margin pressures and are looking for technology to help drive breakthroughs in patient outcomes, and improve operational and clinical productivity by finding new areas of growth. Across the health ecosystem, it is becoming clear that AI and analytics will be the game changers that will provide positive outcomes for patients, providers, and caregivers. One remarkable asset of the health ecosystem is the availability of data because of digitization of patient records, imaging data, and availability of genomic information which needs to be aggregated to analytics and deep learning (DL) algorithms in order to improve results, reduce inefficiencies, and eliminate costly and harmful errors.

Figure 8.1 shows the Big Data using AI that can be stored through electronic health records (EHRs) or precision medicine platforms and shared through cloud systems for the analysis of the data by physicians or researchers that enable the precision medicine for various diseases [33].

The (EHRs) aid in the process of treatment and diagnosis as they are evaluated by ML for each individual [33]. ML is merely an assistant or an augmenter in medical care and is not intended to replace human physicians by any means. The number of radiologists remains the same, while the radiology scans increase continuously. In this case, the use of AI can reduce the time taken by interpreting the data faster alongside its working capacities around the clock. Additionally, the software is not

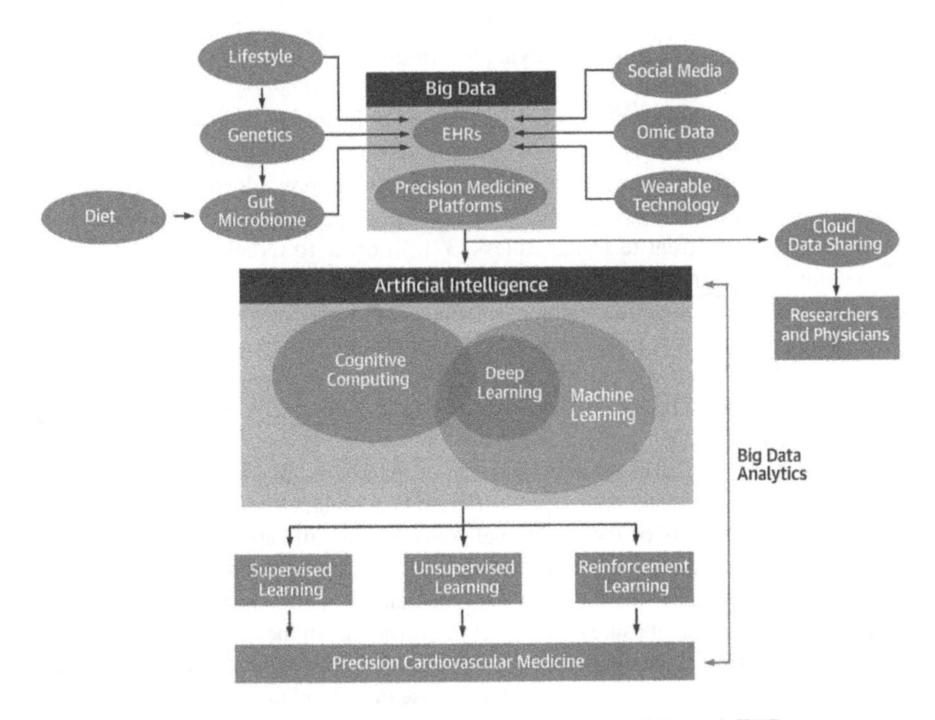

FIGURE 8.1 Big Data using artificial intelligence can be stored through EHRs.

obstructed by human weaknesses such as environmental distractions or interruptions and fatigue that may hinder the performance by reducing its accuracy. In pathology, AI-integrated systems can assist in countering the increasing workload that is followed by an insufficient manpower. Tedious tasks such as the evaluation of morphology and quantitative analysis can also be performed using these systems [34]. Nonetheless, pathologists could utilize the evaluations done by using AI due to its advantages such as reproducible results and decreased variability while diagnosing diseases. This leads to the prevention of misdiagnosis, thereby promoting an earlier detection and diagnosis by alerting the physician to a specific field.

Notably, AI techniques developed in nonmedical areas can be applied for discovering drugs for novel coronavirus. However, for containment, AI can be more than just experimental. Cross-infection is very real for this highly contagious disease, and so the medical workers are highly vulnerable with sheer shortage of protective gear. AI machines are very useful in this scenario, where they collect infected linen and medical garbage, and even disinfect themselves. They not only restrict cross-infection but also reduce the workload. Their ability to read maps and plan the most effective routes makes them efficient.

8.2.3 Precision Medicine

At the dawn of the 20th century, the life expectancy rate according to the World Health Organization was 35 years but it nearly doubled towards the end of the century [35]. This possibility could only be achieved due to improved sanitation, therapies and drugs, and the development of preventive medicine like vaccines that aimed to assist and promote healthier lifestyle [36]. Intuition medicine was practiced in the past where the treatments were according to the knowledge possessed by physician when symptoms of the patient were presented. Evidence-based medicine is quite popular in medical research in which scientific research and appreciable clinical trials have been performed. As the evolution in medicine prevails, we are paying the price for an increase in life expectancy by dedicated approximately 50% of our budget to healthcare to treat terminal illness [37]. In order to target such issues in the 21st century, medicines must focus on accomplishing the four Ps stated by Dr. Leroy Hood: prediction, prevention, personalization, and participation [38] to undergo a transition from traditional medicine to a system that targets the disease with a more personalized approach. The characteristic traits or genomes of a patient, including their lifestyle, will be taken into consideration, and medicine is practiced according to these algorithms [39].

Researchers have learnt from studies that despite being diagnosed with the same disease, these diseases look completely different on a molecular level for such individuals. Present treatments or the traditional "one-size-fits-all" approaches for asthma, diabetes, and other diseases are based on obvious symptoms, standard tests, and common drugs. The computational advances in the field of biology for analyzing the data have created a convenient route to interpret the variations in the disease-specific genotype. This medicine assists in providing personalized prevention and treatment plans to improve a targeted care for a particular disease or individual.

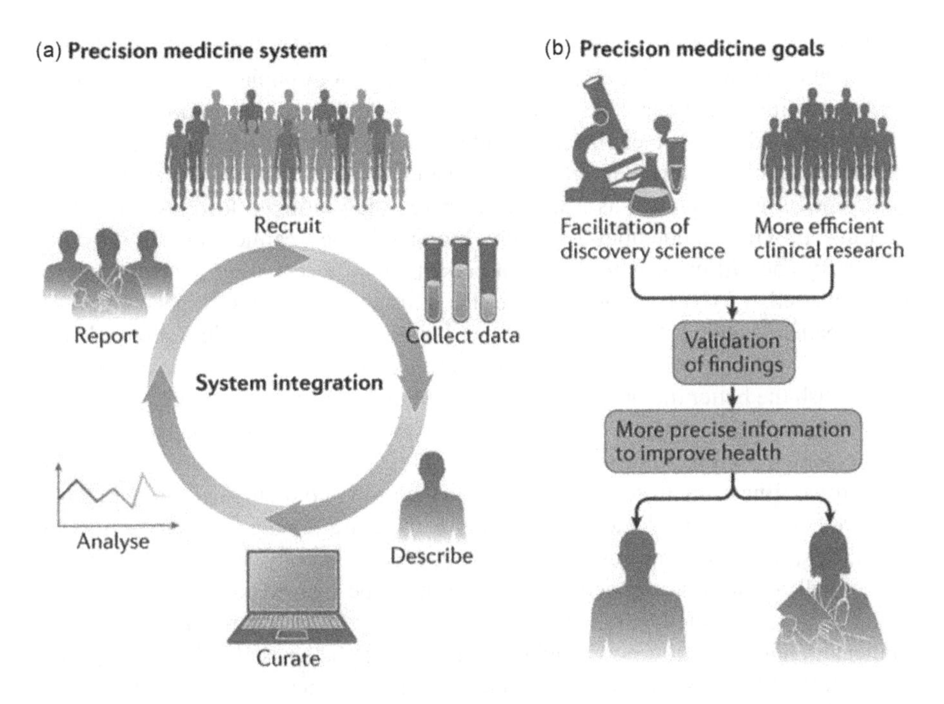

FIGURE 8.2 Essential features and goals of a precision medicine system [41].

The Precision Medicine Initiative (PMI) [40] defined *Precision Medicine* as "an innovative approach that takes into account individual differences in people's genes, environments, and lifestyles." *Precision medicine* is based on collecting, connecting, and analyzing the different kinds of information concerning an individual to better understand health (Figure 8.2).

It provides an opportunity to comprehend diseases in a radically different approach, which may be more effective, intending to develop tailored interventions considering the difference between various people and their conditions [42]. This facilitates the development of new drugs and increases the understanding between the interactions of genomics and response of these drugs, aiming to address different health challenges. The modern approach to understanding health modalities has been achieved by giving the right treatment to the right patient at the right time [43]. The possibility of a more efficient use of social indicators linked to a patient adopts a more holistic approach where relationships and certain experiences have a remarkable impact on the health and well-being of a person [44].

8.2.3.1 Applications of Precision Medicine

Over the past several decades, researches have been conducted continuously at an increasing pace involving the mechanisms in the pathogenesis of cancer and understanding oncogenes as the mortality rate rises worldwide due to cancer. Henceforth, the discovery and development of drugs to suppress the expression of these oncogenes by inactivating their pathways is one of the most favorable features in precision medicine as predicted by the World Health Organization [45]. The clinically applied treatment of chronic myeloid leukemia (CML) is an apt example for this concept.

Another important contribution with innumerable possibilities in precision medicine is the field of tissue engineering and regenerative medicine that comprises the two main topics of interest: organ-on-a-chip and stem cell therapy. Organ-on-a-chip has gained a great interest as it has been increasingly utilized for preclinical tests of drugs, proving its potential use in precision medicine treatment by studying how these drugs act differently on tissues of patients by differentiating induced pluripotent stem cells (iPCs) from adult tissue like skin or blood to the target tissue [46]. These patient-specific 3D tissue-engineered systems with microfluid channels can be used to study any substance virtually and in vitro prior to the clinical treatment to establish the better therapeutic approach to treat a particular disease. Personalized stem cells used to regenerate or replace tissues or organs also focus on the customizable approach used in precision medicine.

Chronic illness seems to be increasing as the growth of complexity in patients is observed due to several factors. For example, there are hundreds of different types of cancers, and each one responds to treatments differently as every human and cancer cell has a unique genetic profile that must be tackled in a specific pattern. Additionally, vast amounts of data or information is required in regard to the development of algorithms to foresee and treat diseases on the basis of individual-specific variabilities like lifestyle, genetics, their response to drugs, and so on. Hence, an alternative to performing an independent research is by launching initiatives to integrate and compile data from various medical units. The collection and analysis of several databases aiming to create new algorithms with the help of tools that are currently available like cloud computing, Big Data analytics, and AI will improve and promote a direct clinical practice. Several issues that arise can be overcome by creating computer-based approaches: analytical methods and AI. Health intelligence "uses tools and methods from AI and data science to provide better insights, reduce waste and wait time, and increase speed, service efficiencies, level of accuracy, and productivity in health care and medicine" [47]. The PMI announced by the National Institute of Health (NIH) in 2015 is a program to deliver resources that aim to enhance healthcare, improve and increase its effectiveness by utilizing the cutting-edge technologies and interdisciplinary research in various fields of medicine [48]. Big Data analytics (genetics, environmental, and lifestyle-related factors) can be shared for data analysis that can be utilized by researchers or physicians through cloud systems that are secure and safely stored through the precision medicine platforms. This data analytics uses AI techniques, including ML, DL, or cognitive computing.

8.2.4 Deep Learning

Inventors wondered whether the programmable computers, when first launched, were able to think or possess the cognitive abilities to learn, understand, and form concepts by applying logic and reason. AI is a technique used for incorporating human intelligence into machines, a burgeoning field in which several active researches are being done alongside its many practical applications. This intelligent software was developed in order to extend automating production, understand and formulate patterns by assisting in the diagnosis and treatment of diseases in medicine by supporting the basic scientific research. Additionally, AI was aimed at tackling rational and complex problems considering different dimensions as humans showed incompetency for such tasks. The solution to these intuitive problems can be achieved by allowing machines to learn from vast amounts of data to gain experience and understand the environment with regard to hierarchy of concepts that enables the processing of complicated conceptions through simpler ones avoiding the need for human operators to intervene and specify the knowledge required by the machine. Ironically, abstract tasks are the most difficult for human beings but prove to be simple for a computer. If one draws a graph illustrating how these concepts are built, the graph tends to be deep and is composed of many layers. Hence, this type of approach is termed as *Deep Learning* [49], and AI is a superset of DL and ML.

DL is a type of ML elucidated by models with numerous hierarchical layers of information compared to the conventional "shallow" learning, and has drawn heavily our knowledge on human brain, statistics, and applied math over the past few years. ML comprises algorithms and statistical models used by machines to perform tasks and gather information by improving with experience, whereas DL is a class of ML that uses artificial neural network which automatically processes and learns hierarchical features of concepts or data comprising multiple layers composed of nonlinear and simple modules, which aids in the transformation of the data to depict an important information required to discriminate these data [50]. Both shallow learning and DL are subsets of AI and are included in ML in which learning occurs without programming. Both these methods "capture patterns from complicated, multidimensional raw data and utilize them as discriminative features of data" [51].

DL has emerged as a trending research in correlation with AI. There has been a tremendous growth and popularity due to the evolution of computers to a more powerful machine with large datasets and techniques to train networks. DL has the advantage of learning complicated patterns from raw data in virtue of multiple layers and transformations, whereas the conventional shallow learning comprising a few layers requires handcrafted features drawn out from the data. Irrespective of the various layers, DL approaches can be used in both supervised and unsupervised learning applications [52]. The supervised method has been applied in the information learned in the past to predict one or more labels or outcomes linked with each data point. By the training process, the parameters are adjusted to precisely predict the target. In contrast, the unsupervised or "exploratory" application has been used when the information used to train is neither classified nor labeled where the goal is to summarize, explain, or identify significant patterns in a dataset (as clustering). In fact, DL may combine both of these steps to construct features tuned to a specific problem and combine them into a "predictor" when adequate data is available and labeled [53]. Representations extracted from unsupervised learning can be used for other supervised tasks.

DL typically represents deep neural networks that mimic the functioning of the human brain composed of neurons (send and receive electrochemical signals) using several layers of artificial network of neurons that can generate an automated response or predictions from the input. This approach was derived from research on artificial neurons that was proposed in 1943 as a model for neuron in our brain that assists in processing information [54]. In neural networks, the input layer contains the inputs fed into the system directing it towards a series of multiple hidden layers, which eventually leads to an output layer that presents the output. A layer comprises features or nodes that are connected to the immediately earlier and deeper layers. These nodes in the input layer contain variables that are to be measured in a particular dataset. Multiple hidden layers learning in which the processing is done, exists in a neural network used for DL wherein each layer is responsible for the feature construction of previous layers while contributing to the refinement of deeper layers through training processes. Consequently, the features of algorithms can automatically be engineered and customized according to one or more tasks. These neural network algorithms can be recurrent neural network (RNN), convolutional neural network (CNN), and deep neural network. The dominant ones in image and speech recognition are RNN (allows the neural network to retain memory over time or sequential inputs) and convolutional network, which have proven effective in various ML jobs such as language translation and image captioning [55]. The diverse opportunities and challenges will effectuate the need for improvement in DL to bring it to new frontiers.

8.2.4.1 Application

DL has evolved as a sought-after application in AI as it is an emerging field and appears promising. It was successfully applied to overcome the challenges posed by ML methods in visual recognition tasks and has led to a dramatic improvement in various scientific and industrial fields such as drug discovery, bioinformatics, and speech recognition [56]. Subsequently, in several image processing and computer vision tasks, DL techniques have surpassed all the available methods due to their flexibility and high accuracy, while the development of new modules enabled the learning of deep conceptions through the efficient use of hardware, including graphic processing units (GPUs). Most importantly, the contribution of large structured datasets such as ImageNet has facilitated the success of this technique [57].

This new ML technique plays a powerful role in image recognition (facial recognitions and image search in Google), self-driving cars, speech recognition (Apple's Siri, Google assistant, Amazon's Alexa), mobile applications (Cardiogram app), and machine vision software in cameras and robots. Furthermore, it can also be used for unsupervised learning tasks to detect novel drug–drug interaction by training it in an unsupervised manner with no limitations on working memory. These algorithms will assist the artificial real-time imaging with improved resolution, thereby reducing cost and enhancing the quality of care. DL AI utilizes Big Data for pattern recognition in heterogeneous syndromes and image recognition. The implementation of DL for Big Data analytics with its unsupervised features holds a remarkable potential to identify the novel phenotypes and genotypes. Big Data can generate the automated hypotheses instead of physicians having to initiate them and henceforth augment in making better clinical decisions instead of replacing the physicians [58]. Tasks can be performed without human intervention in industry such as learning math to write math textbooks, reading publications to answer scientific questions, or watching movies to answer questions. Several indispensable advances and new techniques have created a surge in this field and have enabled their application in large datasets which are sufficient to fit a wide range of parameters that exist for neural networks [59].

8.2.4.2 Implementation of Deep Learning in Medicine

There has been a rapid increase in biomedical data that includes medical records and images along with omics data over the years. Nonetheless, the complex, heterogonous, and multidimensional nature of these data makes it laborious to interpret its clinical meaning. DL permits computational models composed of numerous processing or hidden layers "to learn representations of data with multiple levels of abstraction" [60]. An application of AI can be adapted which is expanding and extensively applied that provides the system with the ability to capture patterns from the biomedical data and employ them in decision-making without being explicitly programmed.

Data-rich disciplines such as biology and medicine are complex in terms of information and are often misinterpreted. Therefore, we employ DL in different applications to a variety of biomedical problems; fundamental biological processes and treatment; and patient classification. This type of learning shows an increased flexibility over other approaches that require large training datasets in order to insert hidden layers as well as to label accurately for supervised learning applications, and due to the aforementioned reasons, DL has become popular particularly in areas of biology and medicine (healthcare and drug discovery) with lower adoption in others. Creation of DL systems assists clinicians and biologists to streamline tasks that do not require experts' advice and prioritize experiments. It can also be applied to answer fundamental biological questions, including the recognition of functional genomic elements such as enhancers and promoters or harmful effects of nucleotide polymorphism [61], and utilizing the deep RNNs to predict gene targets of miRNA and CNNs to predict the protein residue–residue contacts and secondary structure [62].

Accurate classification and precise knowledge of disease and disease subtypes constitute a key challenge in biomedicine; in oncology, current approaches such as histology demand the interpretations by experts or assessment of biomolecular markers or gene expressions that prove to be laborious and critical. An example is the PAM50 approach to classify breast cancer; these patients are divided into four subtypes depending on the expression of 50 marker genes [63]. DL methods have been used in several studies to better classify patients with the help of an unsupervised approach [64]. These automated algorithms play a crucial role and have delivered impressive results across various domains by extracting the meaningful patterns, leading to actionable knowledge by revolutionizing how we categorize patients depending on their diseases and develop treatments within confidential environment but a number of challenges exist on the integration of molecular and imaging data with other types such as EHRs. Comparative study suggests that within the next decade, genomics will surpass all the other fields in data generation and analysis, and such complexities pose as a challenge despite presenting new opportunities [65]. Some efforts aim to identify drug targets and interactions, or recognize and predict drug response and to predict drug bioactivity; some use DL of protein structures [66].

Implementation of DL in medicine aims to include the following: first, unsupervised DL may promote the inspection of novel factors or add the hidden risk factors to present models. Second, it can classify novel genotypes or phenotypes of heart failure with novel diagnostic echocardiographic parameters potentially leading to targeted therapy. Finally, these automated prediction models may be used to predict the risk of bleeding and stroke, and may also help to identify those risk factors, thus incorporating it into new models for therapy. Many multinational technology companies like Apple, Google, and IBM are investing in order to facilitate the development of precision medicine in healthcare analytics. It is important to note that physicians possess the knowledge on how to use AI adequately to perform data analytics, optimize AI applications in clinical diagnosis to accelerate precision medicine, and generate hypotheses. Cognitive computing, ML, and DL can potentially enhance and improve the way medicine is practiced alongside facing the challenges it may pose leading to a greater impact.

8.2.4.3 Convolutional Neural Networks

Evidence has been demonstrated on the efficiency of the CNN methods in terms of accuracy in diagnosis that stands at par with medical practitioners. When each layer of the CNN has filters to go through the height and width of the layers, it is a particular color or the edge of an item that could activate the network. Handwriting patterns of patients with Parkinson's disease were studied using the CNN methods [67], a study with time-series information of healthy and Parkinson's patients.

Neural networks work in a way that neurons receive inputs; ConvNet also works in a similar way, but the architecture assumes it accepts images as inputs. This ConvNet (Figure 8.3) architecture in CNNs is widely used now to understand that the inputs are images [68].

While the method of crowdsourcing could be used to study larger datasets, there are state-of-the-art techniques in DL methods that provide a focus on improved accuracy of images of human organs [68,69]; a work by H. R. Roth et al. has gained popularity on pancreatic images with a proposal on their ConvNet architecture (Figure 8.4).

Image classifiers with CNN were developed to perform the skin cancer classification with least cost-involved experiments [70], and there was an effort to outperform all the CNN tests developed with other datasets until then in the field. Most studies in CNN have taken place in diagnosing breast cancer, which is the most common type of cancer with benign/malignant images fed from raw pixels [71]. Monochromatic images were converted to colorful images [72] for learning breast cancer image datasets while using the popular VGG16 architecture [73], which could handle large depth images using the 3×3 convolution filters and only color images. Thus, there has been a lot of studies, and still it needs further research in cancer diagnosis using images of the CNN patterns.

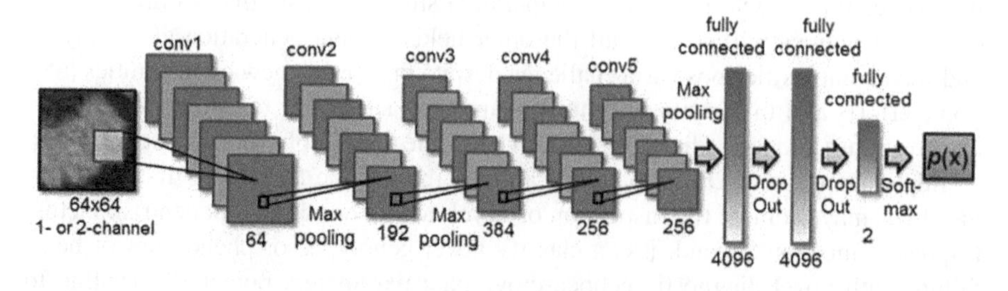

FIGURE 8.3 CNN image filters and neural network connections [68].

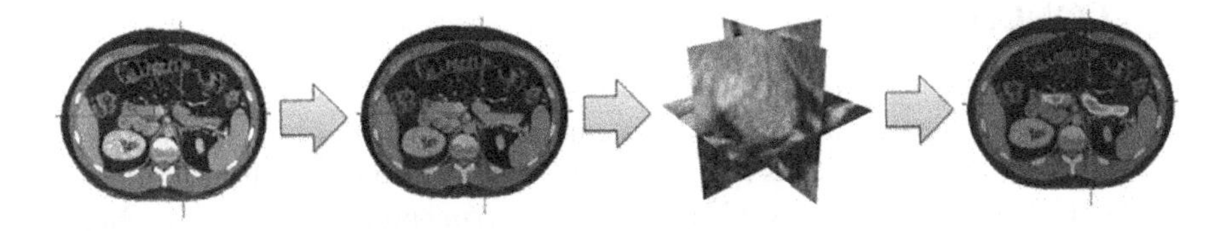

FIGURE 8.4 Axial CT slice of a manual (gold standard) segmentation of the pancreas using CNN [68].

8.2.4.4 CNN in Precision Medicine

While CNN extends its protein region diagnosis [74] to oncology image recognition, expert systems are experimented with a complete diagnosis on the body of the patient to find cancerous cells and treatment plans, which are proposed thereafter by intelligent systems [75]. Mostly, the noise and complexity [76] of the data drugs under consideration causes a false-positive analysis during predictions in precision medicines. These concepts help in metabolomics [76] and studies of organisms while considering their importance in the accurate precision medicine. This is where more accurate and weighted techniques of neural networks come to play a predominant role in filtering images of molecules of drugs to predict their effect on human body organs with conditions especially in oncological precision medicines.

Researchers claimed that effective predictions and findings can be made from chemical structures alone. Studies handled both 2D drug images fed to CNN and molecular fingerprints to Random Forests [77], which are the popular DL techniques. The results proved that the effectiveness of chemical structures through these methods determines drug functions and their mechanism classes [66]. Transcriptomic data of drugs were the input for applying to the CNN patterns to understand and predict drug properties that could be used to treat different systems of the body such as nervous system, cardiology system, or antineoplastic agent [66], thus acting as a precision medicine. Such systems, as shown in Figure 8.5, could predict the functional class, efficacy of the drug, therapeutic use, toxicity of the medicine on human body etc. These techniques in CNN and expert systems have been found to be more advantageous in cancer treatments when traditional endoscopic techniques have their own shortcomings in early diagnosis and treatments [78].

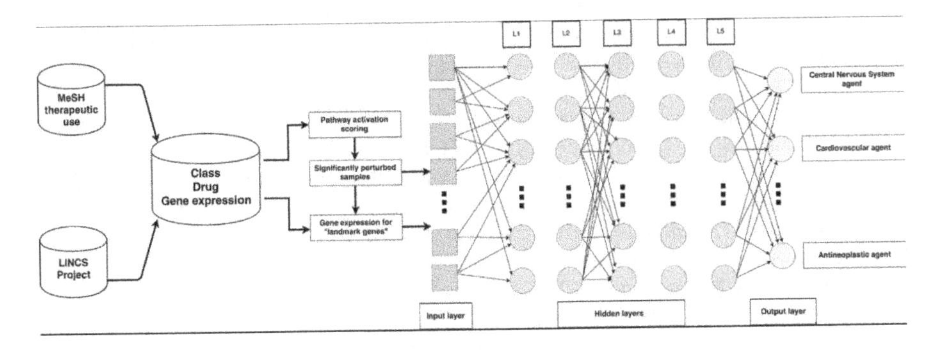

FIGURE 8.5 Neural nodes trained with data from MeSH and LINCS projects [66].

8.5 CONCLUSION

AI and DL techniques have contributed much to the medical field, especially in the field of cancer diagnosis and precision medicine. Major developments were noted with the evolution of techniques like CNNs in DL. Research studies have established a greater platform with AI and CNN techniques entering precision medicines for personalized treatments. These studies have given a quintessential support to many of the existing techniques to obscure imaging credibility in the field of oncology. These studies include diagnosing cancer with imaging techniques using AI to understand the affected areas of human body with more specificity. Popular AI techniques like DL, namely, CNN and Random Forests, play further profound roles in understanding the molecular structure of cancer drugs to evaluate their effects on cancer cells, thus assisting in precision medicines.

Studies have also explored the interconnection of the brain and the central nervous system with the aid of nanotechnology [79,80]. There has been a drastic advancement in the field of medicine with the integration of ML and AI with computer interfaces using nanotechnology [81]. As the convergence of ML and nano-engineered medical components has not taken place yet, the establishment of expert systems and structured role in several aspects of nanotechnology and molecular-level drug research [82] can still bring an arsenal of new methodologies and frameworks to wade away cancer at a very early stage with the precision medicine before it infests other human cells.

REFERENCES

1. K. Rajasundari and K. Ilamurugu, "Nanotechnology and its applications in medical diagnosis," *Journal of Basic and Applied Chemistry*, vol. 1, no. 2, pp. 26–32, 2011.
2. H. Laroui, P. Rakhya, B. Xiao, E. Viennois, and D. Merlin, "Nanotechnology in diagnostics and therapeutics for gastrointestinal disorders," *Digestive and Liver Disease: Official Journal of the Italian Society of Gastroenterology and the Italian Association for the Study of the Liver*, vol. 45, no. 12, pp. 995–1002, Dec. 2013.
3. K. B. Sutradhar and M. L. Amin, "Nanotechnology in cancer drug delivery and selective targeting," *ISRN Nanotechnology*, vol. 2014, pp. 1–12, 2014.
4. X. Luan, K. Sansanaphongpricha, I. Myers, H. Chen, H. Yuan, and D. Sun, "Engineering exosomes as refined biological nanoplatforms for drug delivery," *Acta Pharmacologica Sinica*, vol. 38, no. 6, pp. 754–763, Jun. 2017.
5. J. M. Lewis, A. D. Vyas, Y. Qiu, K. S. Messer, R. White, and M. J. Heller, "Integrated analysis of exosomal protein biomarkers on alternating current electrokinetic chips enables rapid detection of pancreatic cancer in patient blood," *ACS Nano*, vol. 12, no. 4, pp. 3311–3320, Mar. 2018.
6. A. Arima, M. Tsutsui, I. H. Harlisa, T. Yoshida, M. Tanaka, K. Yokota, W. Tonomura, M. Taniguchi, M. Okochi, T. Washio, and T. Kawai, "Selective detections of single-viruses using solid-state nanopores," *Scientific Reports*, vol. 8, no. 1, pp. 1–7, Nov. 2018.
7. L. Bergeron, "Stanford researchers' magnetic nanotags spot cancer in mice earlier than current methods," *Stanford University*, 13-Oct-2009. [Online]. Available: https://news.stanford.edu/news/2009/october12/cancer-detection-101209.html. [Accessed: 04-Apr-2020].
8. T. L. Halo, K.M. McMahon, N.L. Angeloni, Y. Xu, W. Wang, A.B. Chinen, D., Malin, E. Strekalova, V.L. Cryns, C. Cheng, and C.A. Mirkin, "NanoFlares for the detection, isolation, and culture of live tumor cells from human blood," *Proceedings of the National Academy of Sciences*, vol. 111, no. 48, pp. 17104–17109, Dec. 2014.

9. R. Ps, B. We, C. Ab, G. Cm, P. Sh, and M. Ca, "NanoFlares as probes for cancer diagnostics," *Cancer Treatment and Research*, 2015. [Online]. Available: https://pubmed.ncbi.nlm.nih.gov/25895862/. [Accessed: 06-Apr-2020].
10. T. L. Halo et al., "NanoFlares for the detection, isolation, and culture of live tumor cells from human blood," *Proceedings of the National Academy of Sciences*, vol. 111, no. 48, pp. 17104–17109, Dec. 2014.
11. R. Marega et al., "Antibody-functionalized polymer-coated gold nanoparticles targeting cancer cells: an in vitro and in vivo study," *Journal of Materials Chemistry*, vol. 22, no. 39, p. 21305, 2012.
12. Researchers, "UCLA researchers further refine 'NanoVelcro' device to grab single cancer cells from blood," *UCLA*. Available from: [https://www.ncbi.nlm.nih.gov/pmc/articles/PMC4924510/.
13. S. Chen, S. K. Boda, S. K. Batra, X. Li, and J. Xie, "Emerging roles of electrospun nanofibers in cancer research," *Advanced Healthcare Materials*, vol. 7, no. 6, p. e1701024, Mar. 2018.
14. Y.-E. Choi, J.-W. Kwak, and J. W. Park, "Nanotechnology for early cancer detection," *Sensors*, vol. 10, no. 1, pp. 428–455, Jan. 2010.
15. Z. Chen et al., "The advances of carbon nanotubes in cancer diagnostics and therapeutics," *Journal of Nanomaterials*, vol. 2017, pp. 1–13, 2017.
16. T. Yasui et al., "Unveiling massive numbers of cancer-related urinary-microRNA candidates via nanowires," *Science Advances*, vol. 3, no. 12, p. e1701133, Dec. 2017.
17. S. Fahmy, "UGA researchers develop rapid diagnostic test for pathogens, contaminants," *UGA Today*, 19-Jul-2012. Available from: https://www.qualityassurancemag.com/article/uga-researchers-develop-rapid-diagnostic-test/
18. C. R. Rekha, V. U. Nayar, and K. G. Gopchandran, "Synthesis of highly stable silver nanorods and their application as SERS substrates," *Journal of Science: Advanced Materials and Devices*, vol. 3, no. 2, pp. 196–205, Jun. 2018.
19. A. B. Chinen, C. M. Guan, J. R. Ferrer, S. N. Barnaby, T. J. Merkel, and C. A. Mirkin, "Nanoparticle probes for the detection of cancer biomarkers, cells, and tissues by fluorescence," *Chemical Reviews*, vol. 115, no. 19, pp. 10530–10574, Aug. 2015.
20. Z. R. Stephen, F. M. Kievit, and M. Zhang, "Magnetite nanoparticles for medical MR imaging," *Materials Today (Kidlington, England)*, vol. 14, nos. 7–8, pp. 330–338, 2011.
21. Researchers, "News," *CNS Oncology*, vol. 2, no. 1, pp. 7–10, Jan. 2013.
22. M. Fang, C. Peng, D.-W. Pang, and Y. Li, "Quantum dots for cancer research: current status, remaining issues, and future perspectives," *Cancer Biology & Medicine*, vol. 9, no. 3, pp. 151–163, Sep. 2012.
23. A. B. de González, "Projected cancer risks from computed tomographic scans performed in the United States in 2007," *Archives of Internal Medicine*, vol. 169, no. 22, p. 2071, Dec. 2009.
24. S. Singh, M. Nagpal, P. Singh, P. Chauhan, and M. Zaidi, "Tumor markers: a diagnostic tool," *National Journal of Maxillofacial Surgery*, vol. 7, no. 1, p. 17, 2016.
25. Anonymous, "Tumor Markers," *National Cancer Institute*, 06-May-2019. Available from: https://www.cancer.gov/about-cancer/diagnosis-staging/diagnosis/tumor-markers-fact-sheet
26. T. Malati, "Tumour markers: an overview," *Indian Journal of Clinical Biochemistry*, vol. 22, no. 2, pp. 17–31, Sep. 2007.
27. J. V. Frangioni, "New technologies for human cancer imaging," *Journal of Clinical Oncology*, vol. 26, no. 24, pp. 4012–4021, Aug. 2008.
28. G. C. Bauer and B. Wendeberg, "External counting of Ca47 and Sr85 in studies of localised skeletal lesions in man," *The Journal of Bone and Joint Surgery. British Volume*, vol. 41-B, pp. 558–580, Aug. 1959.

29. P. C. Bailey and S. S. Martin, "Insights on CTC biology and clinical impact emerging from advances in capture technology," *Cells*, vol. 8, no. 6, Jun. 2019.
30. A. Kaplan and M. Haenlein, "Siri, Siri, in my hand: Who's the fairest in the land? On the interpretations, illustrations and implications of artificial intelligence," *Business Horizons*, vol. 62, no. 1, pp. 15–25, 01 Jan. 2019.
31. A. Turing, "Computing machinery and intelligence," Mind, vol. 49, pp. 433–460, 1950.
32. H. Yokota, M. Goto, C. Bamba et al., "Reading efficiency can be improved by minor modification of assigned duties: a pilot study on a small team of general radiologists," *Japanese Journal of Radiology*, vol. 35, pp. 262–268, 2017.
33. C. Krittanawong, H. Zhang, Z. Wang, M. Aydar, and T. Kitai, "Artificial intelligence in precision cardiovascular medicine," *Journal of the American College of Cardiology*, vol. 69, no. 21, pp. 2657–2664, 2017.
34. Y. Mintz and R. Brodie, "Introduction to artificial intelligence in medicine," *Minimally Invasive Therapy & Allied Technologies*, vol. 28, pp. 1–9, 2019.
35. J. C. Riley, "Estimates of regional and global life expectancy, 1800–2001," *Population and Development Review*, vol. 31, no. 3, pp. 537–543, 2005. doi: 10.1111/j.1728-4457.2005.00083.x.
36. D. Cutler, A. Deaton, and A. Lleras-Muney, "The determinants of mortality," *Journal of Economic Perspectives*, vol. 20, no. 3, pp. 97–120, 2006. doi: 10.1257/jep.20.3.97.
37. L. E. Garcez-Leme and M. D. Leme, "Costs of elderly healthcare in Brazil: challengers and strategies," *MedicalExpress*, vol. 1, no. 1, pp. 3–8, 2014.
38. D. J. Galas and L. Hood, "Systems biology and emerging technologies will catalyze the transition from reactive medicine to predictive, personalized, preventive and participatory (P4) medicine," *Interdisciplinary Bio Central*, vol. 1, no. 2, pp. 1–4, 2009.
39. G. Gameiro, V. Sinkunas, G. Liguori, and J. Auler-Júnior, "Precision medicine: changing the way we think about healthcare," *Clinics*, vol. 73, p. e723, 2018.
40. The precision medicine initiative. June 2020. Accessed on https://obamawhitehouse.archives.gov/precision-medicine.
42. S. A. Dugger, A. Platt, and D. B. Goldstein, "Drug development in the era of precision medicine," *Nature Reviews Drug Discovery*, vol. 17, no. 3, pp. 183–196, 2018.
43. E. K. Shin, R. Mahajan, O. Akbilgic, and A. ShabanNejad, "Sociomarkers and biomarkers: predictive modeling in identifying pediatric asthma patients at risk of hospital revisits," *npj Digital Medicine*, vol. 1, p. 50, 2018.
44. J. H. Brenas, E. K. Shin, and A. ShabanNejad, "Adverse childhood experiences ontology for mental health surveillance, research, and evaluation: advanced knowledge representation and semantic web techniques," *JMIR Mental Health*, vol. 6, no. 5, p. e13498, 2019.
41. E. M. Antman and J. Loscalzo, "Precision medicine in cardiology." *Nature Reviews Cardiology*, vol. 13, no. 10, pp. 591–602, 2016.
45. WHO, Cancer. *World Health Organization*, 2018. Available from: http://www.who.int/mediacentre/factsheets/fs297/en/. [Accessed: 10-Feb.-2018].
46. A. Williamson, S. Singh, U. Fernekorn, and A. Schober, "The future of the patient-specific body-on-a-chip," *Lab on a Chip*, vol. 13, no. 18, pp. 3471–3480, 2013.
47. A. ShabanNejad, M. Michalowski, and D. L. Buckeridge, "Health intelligence: how artificial intelligence transforms population and personalized health," *npj Digital Medicine*, vol. 1, no. 53, 2018.
48. J. Steenhuysen, "NIH director sees solving data puzzle as key to U.S. precision medicine," Reuters, 2015. Available from: https://www.reuters.com/article/us-usa-health-precision/nih-director-sees-solving-data-puzzle-as-key-to-u-s-precision-medicine-idUSKBN0M302520150307.
49. I. Goodfellow, Y. Bengio and A. Courville, *Adaptive Computation and Machine Learning Series*. Cambridge, MA: MIT Press, 2017.

50. Y. LeCun, Y. Bengio, and G. Hinton, "Deep learning," *Nature*, vol. 521, pp. 436–444, 2015.
51. H. Choi, "Deep learning in nuclear medicine and molecular imaging: current perspectives and future directions," *Nuclear Medicine and Molecular Imaging*, vol. 52, pp. 109–118, 2018.
52. C. Krittanawong, H. J. Zhang, Z. Wang, M. Aydar, and T. Kitai, "Artificial intelligence in precision cardiovascular medicine," *Journal of the American College of Cardiology*, vol. 69, no. 21, 2017. ISSN 0735-1097.
53. T. Ching, D. S. Himmelstein, B. K. Beaulieu-Jones, A. A. Kalinin, B. T. Do, G. P. Way, E. Ferrero et al., "Opportunities and obstacles for deep learning in biology and medicine," *Journal of the Royal Society Interface*, vol. 15, no. 141, p. 2017038, 2018.
54. W. S. McCulloch and W. Pitts, "A logical calculus of the ideas immanent in nervous activity," *Bulletin of Mathematical Biophysics*, vol. 5, pp. 115–133, 1943.
55. K. Cho, B. Van Merriënboer, C. Gulcehre et al., "Learning phrase representations using RNN encoder-decoder for statistical machine translation," 2014. arXiv: 1406.1078.
56. O. Russakovsky, J. Deng, H. Su, J. Krause, S. Satheesh, S. Ma et al., "ImageNet large scale visual recognition challenge," *International Journal of Computer Vision*, vol. 115, pp. 211–252, 2015.
57. A. Krizhevsky, I. Sutskever, and G. E. Hinton, "ImageNet classification with deep convolutional neural networks," *Advances in Neural Information Processing Systems*, vol. 25, pp. 1090–1098, 2012.
58. M.-H. Kuo, T. Sahama, A. W. Kushniruk, E. M. Borycki, and D. K. Grunwell, "Health big data analytics: current perspectives, challenges and potential solutions," *International Journal of Big Data Intelligence*, vol. 1, pp. 114–126, 2014.
59. F. Niu, B. Recht, C. Re, and S. J. Wright, "HOGWILD!: a lock-free approach to parallelizing stochastic gradient descent," 2011. arXiv: 1106.5730. https://arxiv.org/abs/1106.5730v2.
60. Y. LeCun, Y. Bengio, and G. Hinton, "Deep learning," *Nature*, vol. 521, pp. 436–444, 2015.
61. D. Quang, Y. Chen, and X. Xie, "DANN: a deep learning approach for annotating the pathogenicity of genetic variants," *Bioinformatics*, vol. 31, pp. 761–763, 2015.
62. S. Wang, S. Sun, Z. Li, R. Zhang, and J. Xu, "Accurate de novo prediction of protein contact map by ultra-deep learning model," *PLoS Computational Biology*, vol. 13, p. e1005324, 2017.
63. I. A. Mayer, V. G. Abramson, B. D. Lehmann, and J. A. Pietenpol, "New strategies for triple-negative breast cancer – deciphering the heterogeneity," *Clinical Cancer Research*, vol. 20, pp. 782–790, 2014.
64. D. C. Cireşan, A. Giusti, L. M. Gambardella, and J. Schmidhuber, "Mitosis detection in breast cancer histology images with deep neural networks." In K. Mori, I. Sakuma, Y. Sato, C. Barillot, and N. Navab (eds) *Medical Image Computing and Computer-Assisted Intervention – MICCAI 2013* (pp. 411–418). Springer, Berlin, 2013.
65. Z. D. Stephens et al., "Big data: astronomical or genomical?" *PLoS Biology*, vol. 13, p. e1002195, 2015.
66. A. Aliper, S. Plis, A. Artemov, A. Ulloa, P. Mamoshina, and A. Zhavoronkov, "Deep learning applications for predicting pharmacological properties of drugs and drug repurposing using transcriptomic data," *Molecular Pharmaceutics*, vol. 13, pp. 2524–2530, 2016.
67. P. Khatamino, İ. Cantürk, and L. Özyılmaz, "A deep learning-CNN based system for medical diagnosis: an application on Parkinson's disease handwriting drawings." In *2018 6th International Conference on Control Engineering & Information Technology (CEIT)*, 2018, October (pp. 1–6). IEEE, Istanbul, Turkey.

68. H. R. Roth, L. Lu, A. Farag, H.-C. Shin, J. Liu, E. B. Turkbey, and R. M. Summers, "Deep organ: multi-level deep convolutional networks for automated pancreas segmentation." In: N. Navab, J. Hornegger, W. M. Wells, and A. F. Frangi (eds) *Medical Image Computing And Computer-Assisted Intervention – MICCAI 2015*, Part I (vol. 9349, pp. 556–564), LNCS, 2015. Springer, Heidelberg.
69. A. Prasoon, K. Petersen, C. Igel, F. Lauze, E. Dam, and M. Nielsen, "Deep feature learning for knee cartilage segmentation using a triplanar convolutional neural network," *Medical Image Computing and Computer-Assisted*, vol. 16, Pt 2, pp. 246–253, 2013.
70. R. Zhang, Y. Zheng, T.W.C. Mak, R. Yu, S. H. Wong, J. Y. Lau, and C. C. Poon, "Automatic detection and classification of colorectal polyps by transferring low-level CNN features from nonmedical domain," *IEEE Journal of Biomedical and Health Informatics*, vol. 21, no. 1, pp. 41–47, 2016.
71. S. Dabeer, M. M. Khan, and S. Islam, "Cancer diagnosis in histopathological image: CNN based approach," *Informatics in Medicine Unlocked*, vol. 16, p. 100231, 2019.
72. Liu, J. et al., "Integrate domain knowledge in training CNN for ultrasonography breast cancer diagnosis," *Lecture Notes in Computer Science (including subseries Lecture Notes in Artificial Intelligence and Lecture Notes in Bioinformatics)*, vol. 11071, pp. 868–875, 2018.
73. K. Simonyan and A. Zisserman, "Very deep convolutional networks for large-scale image recognition," 2014. arXiv preprint arXiv: 1409.1556.
74. S. Wang, S. Weng, J. Ma, and Q. Tang, "DeepCNF-D: predicting protein order/disorder regions by weighted deep convolutional neural fields," *International Journal of Molecular Sciences*, vol. 16, no. 8, pp. 17315–17330, 2015.
75. D. Grapov, J. Fahrmann, K. Wanichthanarak, and S. Khoomrung, "Rise of deep learning for genomic, proteomic, and metabolomic data integration in precision medicine," *Omics: A Journal of Integrative Biology*, vol. 22, no. 10, pp. 630–636, 2018.
76. D. K. Trivedi, K. A. Hollywood, and R. Goodacre, "Metabolomics for the masses: the future of metabolomics in a personalized world," *New Horizons in Translational Medicine*, vol. 3, no. 6, pp. 294–305, 2017.
77. J. G. Meyer, S. Liu, I. J. Miller, J. J. Coon, and A. Gitter, "Learning drug functions from chemical structures with convolutional neural networks and random forests," *Journal of Chemical Information and Modeling*, vol. 59, no. 10, pp. 4438–4449, 2019.
78. M. Ruan, Y. Ren, Z. Wu, J. Wu, S. Wang, and X. Yang, "The survey of CNN-based cancer diagnosis system." In *IOP Conference Series: Materials Science and Engineering* (vol. 466, no. 1, p. 012095), 2018, December.
79. G. A. Silva, "Neuroscience nanotechnology: progress, challenges, and opportunities," *Nature Reviews Neuroscience*, vol. 7, pp. 65–74, 2006.
80. N. A. Kotov, J. O. Winter, I. P. Clements, E. Jan, P. T. Timko, S. Campidelli et al., "Nanomaterials for neural interfaces," *Advanced Materials*, vol. 21, pp. 1–35, 2009.
81. G. A. Silva, "A new frontier: the convergence of nanotechnology, brain machine interfaces, and artificial intelligence," *Frontiers in Neuroscience*, vol. 12, Article 843, Nov. 2018.
82. G. M. Sacha and P. Varona, "Artificial intelligence in nanotechnology," *Nanotechnology*, vol. 24, p. 452002, 2013.

9 Applications of Artificial Intelligence in Pharmaceutical and Drug Formulation

Uddipta Das and Pankaj Wadhwa
Lovely Professional University

CONTENTS

9.1 INTRODUCTION

Artificial intelligence (AI) in the pharma sector utilizes the automated algorithms to achieve the objective, which earlier depended upon human intelligence. The use of AI in the pharma and biotech industry reformulates how new drug evolves, fights against different diseases, and more [1,2]. It is the abstraction of convoluted lore that measures botheration that have their roles in any facet of biological information refine. It is also known as computational intelligence [3–5]. It encompasses three different types: human-created algorithms, machine learning, and deep learning [1]. Artificial neural network (ANN) is one of the best computational tools, which is recognized by scientists and engineers [3,6]. It is inspired from the biological animal

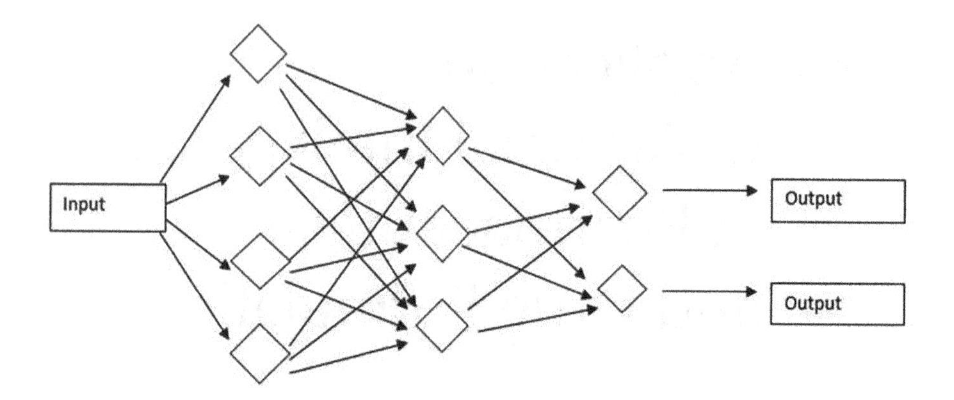

FIGURE 9.1 Working of artificial neural network

networks that constitute animal brains [7,8]. This network makes a resolution and gives outcomes alike when conferred with the fragmentary presentation [9].

ANN resembles the biological nervous system that consists of various processing units and artificial neurons. Its entirety is well for clearing up a nonlinear complication of different components and different response systems such as space analysis in drug discovery [3]. In the pharma sector, it plays an essential role in drug design, drug testing and optimization, and drug interactions. It also helps in finding the optimal formula for the specific drug. As it is stated that preformulation is one of the important steps, the information regarding the property and behavior of some mixtures is available, and feeding this information to the neural network (NN) as an input gives the outcome. By the use of advanced the problem of multiobjective simultaneous optimization, which is incredibly necessary in the case of drugs which combine more than one active substance and the beneficial effects have to be considered in close relationship with adverse side effects can be sorted out [10–12]. The advantages of the NNs are generalizability, flexibility, rapidity, highly sensitivity, and effectiveness. The NN also deals with the complex real-world applications with negligible variations [13]. The general working mechanism of ANN is represented in Figure 9.1.

9.2 GENETIC ALGORITHM

It is an optimization technique and can be defined as stationed on the perception of evolution of biology [13]. It is a one kind of exploration heuristic that basically reproduces or stimulates the action of normal advancement in respect of AI. Its continuous use leads to obtain a better solution to boost and also to explore more complications [14]. It is a vague escalation arrangement and caters a strong way to achieve the focused arbitrary searches in a big complication gap as bump into chemometric and drug development [15]. In this particular aspect, the two major things, namely, genetic representation of the solutions and fitness functions, are to be considered [16]. The main thing to find the solution of any problem is the drawing of genetic depiction. After this genetic depiction, the last one is to classify a fitness function in support of complication. There are peculiar fitness functions for

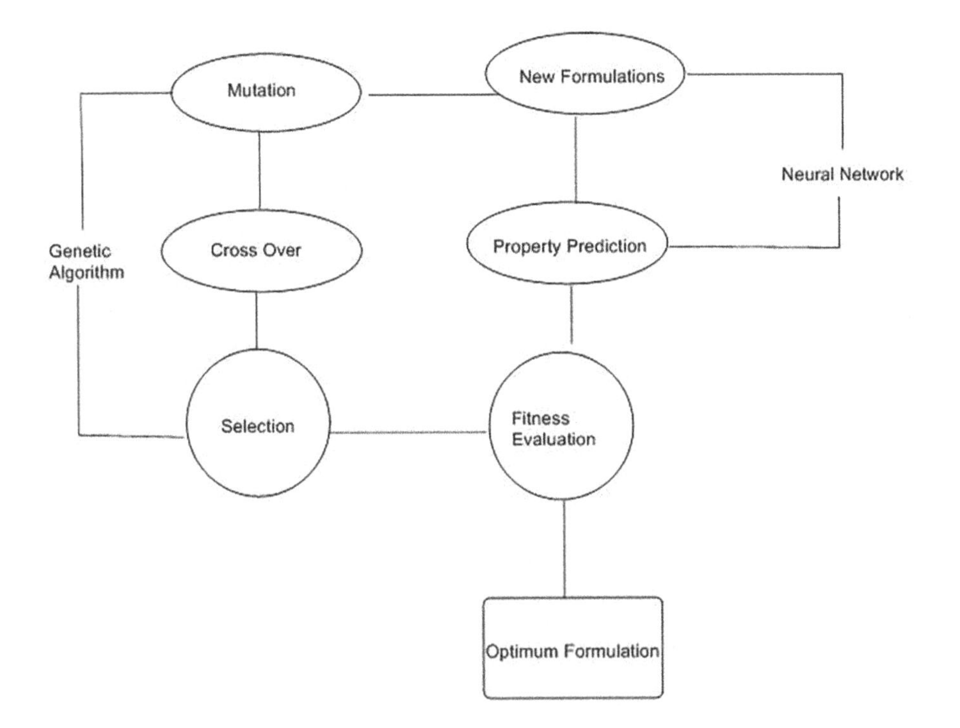

FIGURE 9.2 Working of genetic algorithm.

the different problems. It initiates with drawing of the genetic algorithm (GA) by the way of initialization of the population of solutions that were randomly created to figure out an initial population. The nature of complication always decides the size of the population, and later evaluation of the fitness function will be done for every individual population. In this contrast, the last step is reproduction of population by employing various genetic operator-related functions like selection, crossover, and mutation for the next generation of population [17]. The working principle of the generic algorithm is represented in Figure 9.2.

Here, we have also tried to summarize the link connecting modeling and optimization as shown in Figure 9.3.

In the next part, we will discuss about the concept "fuzzy logic."

9.3 FUZZY LOGIC

Drug discovery is a sensitive, interminable, and successive process that involves the use of a virtual library of compounds along with other computational methods to optimize the lead and target followed by their preclinical studies by the way of *in vitro* and *in vivo* protocols [18]. This technique is helping the scientists over the long period in the area of drug development. It is a science of assumptions that should be similar to human reasoning, and here, all levels of possibilities of input express their outcome in the form of digital values YES and NO [19]. In another words, it can be defined as that its values lie between 0 and 1 along with one term μ, which

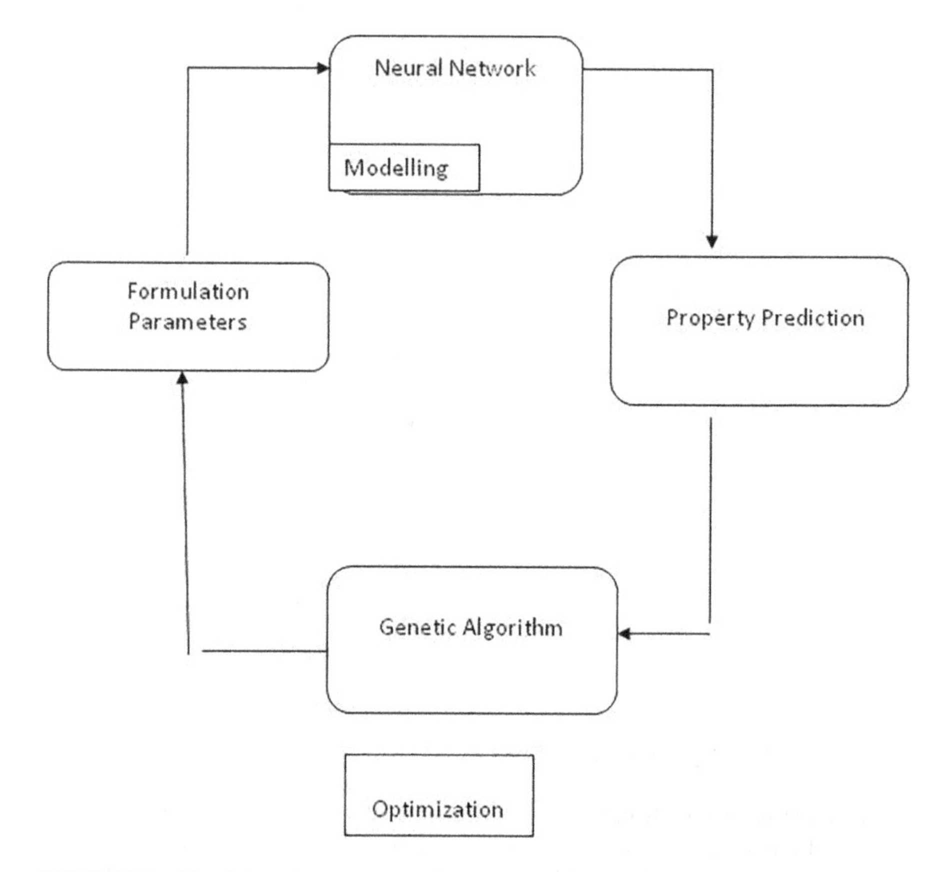

FIGURE 9.3 The liaison between modeling and optimization.

represents the membership function. Fuzzy inference is the most critical characteristic part of this method, which is usually based on the fuzzy set theory, as mentioned in Figure 9.4 [20]. If we talk about their applications, then we can state that these systems have grown vast in consumer electronics and automobiles sectors. The detailed applications will be discussed in the next section.

9.4 INTEGRATED SOFTWARE

Data mining and modeling are the best suitable ways for the above-said but somehow a degree of expertise should be maintained. To be absolutely convenient to the product formulation, these technologies should be integrated into packages that should have essential tools as well as a facility to utilize the rational default values meant for parameters. We can easily correlate by this example; the software package should comprise ANN and GA in case of setting up the optimized formulations. Additionally, the fuzzy logic will create the pragmatic framework for representing the intentions in contrast to optimization in a better way [21]. Nowadays, ANOVA method is usually used for integrating the raw data, and apart from this, other ways like CAD/Chem and IN Form (a product of intelligences) are also used for examining the data [22]. In the next section, we will discuss about various applications of AI and its importance.

9.5 APPLICATIONS OF ARTIFICIAL INTELLIGENCE IN PHARMACEUTICALS

The domain of AI is huge in all directions. We have clearly seen the involvement of AI in the areas of clinical systems, tracking systems, patterns such as face and handwriting followed by speech and voice recognitions. Here, we will discuss the role of AI in the pharmaceutical research.

9.6 RECOGNITION OF PATTERN AND MODELING THE DATA OF ANALYSIS

Pattern recognition is a process of recognizing patterns by using a machine learning algorithm [23]. These NNs are capable of identifying patterns from boisterous and complicated data to appraisal nonlinear relationship, respectively. By using this concept, the ANN can be employed in the field of study of investigation of peak-shaped signals in the obtained analytical data such as various kinds of spectral data. By using this technique, we can speed up the identification of unidentified samples via creating the NN [9,24]. The process that is usually known as conventional multiple linear regression (MLR) for matching the obtained spectra with the known one is little bit lengthy and tricky. To overcome these types of problems, ANN can play a key role in this area [9]. It can be clearly understood by an example of a drug ranitidine hydrochloride that belongs to an antihistaminic class and is available in two different forms, viz, Forms 1 and 2, respectively. The various types of techniques such IR spectroscopic methods, X ray diffraction in combination with ANN are commonly used as an advanced data modeling tool to assess the available rapid, elementary methods for the qualitative and quantitative control of the above-said drug [9,25]. This advanced technology helps in examining the purity of drug by determining its different crystal forms as well as in identifying its polymorphic transition by quantifying it. Apart from this role, the problem associated with a considerable overlap of the tablet ingredients' spectral pattern can be sorted out via ANN [9].

9.7 MODELING THE RESPONSE SURFACE

The benefit of ANN for the response surface modeling for the optimization of HPLC can be achieved side-by-side with nonlinear regression methods. Chromatographic behavior of solutes was explained by retention mapping via observing the response surface that is able to explain the relationship between chromatographic patterns of solutes and mobile-phase components, respectively. By this way, the capacity factor can also be predicted for each solute present in the sample, and it was also reported that the ANN-predicted values are much clear than those acquired by the regression model. Apart from this, chromatographic peak purity by investigating

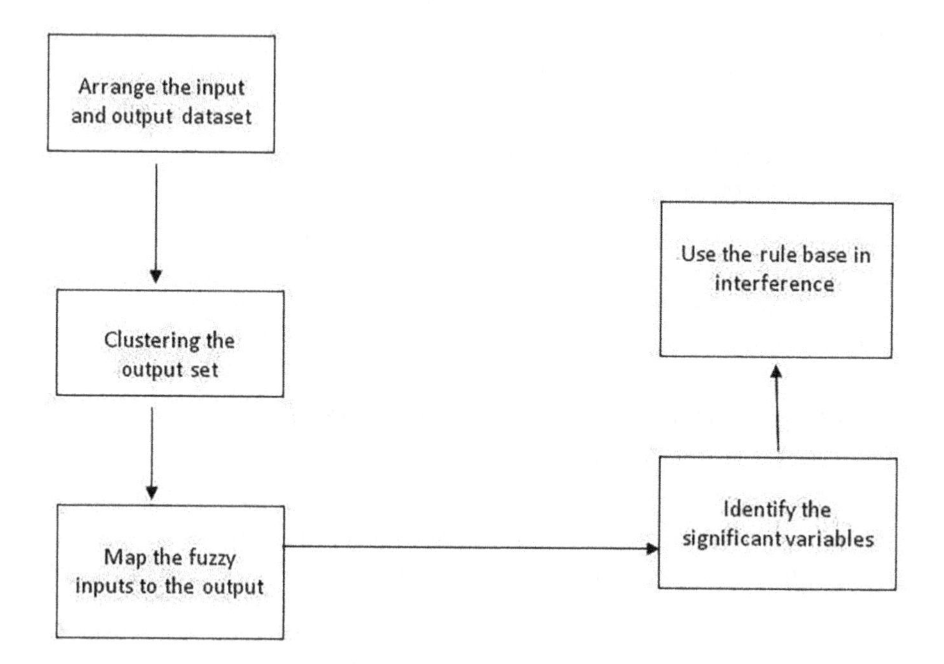

FIGURE 9.4 Steps involved in fuzzy logic.

chromatographic data via the nonlinear transformation function and structure retention relationship from the molecular structures of solutes for predicting their chromatographic can also be assessed [9].

9.8 IN ASSESSMENT OF CONTROLLED-RELEASE AND IMMEDIATE-RELEASE FORMULATIONS

Intended for these kinds of formulations, the scientists have prepared several formulations through ANN and pharmacokinetic (PK) simulations on the basis of formulation variables and other tablet variables as the ANN model inputs. There were ten different sampling time points in which in vitro cumulative percentage of drug released was used as an outfit. CAD/Chem software was used for the advancement of ANN model and skilled for data sets. The prediction of optimal formulation compositions by the trained ANN model was based on *in vitro* dissolution time and *in vivo* release profiles, respectively. The key step was dissolution, and on the basis of that, a conclusion was made that the fraction of the drug absorbed *in vivo* is directly proportional to the dissolution of the drug [1].

Hussain and co-workers from the University of Cincinnati, USA, explored the usage of NNs for modeling the pharmaceutical formulations. In different studies, they have used these methods to determine the in vitro release characteristics of several drugs. The drug release predictions were carried out by NNs through a distinct unseen layer, and the obtained results were comparable with the data that was generated through the statistical analysis. There was no effort to be made to restore the formulation using GAs, but the obtained results have motivated to nominate the computer-aided formulation design established on NN [26,27]. Another study was also performed to predict the drug release rate and to carry out the optimization through 2-D or 3-D response analysis of surface by the way of NN on the basis of the tablet matrix. Their obtained results have shown a nonlinear relationship between the release rate and the amount of ingredients that are used for the production of several formulations [12,18,27]. Another group have also developed an ANN model to optimize a sustained release of diclofenac sodium matrix tablets. The ingredients and sampling time were considered as formulation variables and input, respectively. In the hidden layer, 12 hidden nodes were added. The drug release was used as an output at each sampling point and expressed in the form of percentage. A trained ANN model was engaged to anticipate the release profile and optimize the formulation composition established on the drug released, which was expressed in percentage [28].

Tokayama and their group members have developed a simultaneous optimization technique for optimizing a controlled release of theophylline tablets using cornstrose along with a mixture of hydroxyl propyl methyl cellulose, lactose, and cornstarch. The sum of the slow- and fast-release fraction was characterized by the release profile of theophylline. To build the ANN model, the amount of cornstrose as well as cornstarch and compression pressure were taken as the casual factors; the primary amount of theophylline and its rate constants in the fast- and slow-release fractions were considered as response variables. The ANN model result showed that the theophylline release rate was equal to the absorption rate, and the optimization of the prepared formulations was done by the distance function method [29].

Ibric et al. designed the extended release of aspirin tablets via the generalized regression neural network (GRNN) by considering the amount of Eudragit RS PO and pressure of compression as the casual factors. The *in vitro* dissolution time profiles, coefficients n (release order), and log k (release constant) were noted down as the release parameters, respectively. The obtained model was also able to anticipate the formulation and process factors for *in vitro* drug release profiles. Later on, the same GRNN model was also employed for predicting the drug stability and in vitro–in vivo correlation [30,31]. In the next section, we will discuss about the role of AI in the immediate-release formulations.

In order to maximize the tablet strength on the basis of the selected lubricant, Turkoglu et al. have modeled against hydrochlorothiazide formulation [32]. Similarly, Kesavan and Peck modeled the uses of caffeine tablet formulation to recite both formulation and processing variables, respectively. Based on these observations, it was stated that the NN performs better than the conventional statistical method. The combination of NN and GA can be employed for reinvestigating the data and their results [33].

For different characteristics of product performance, many optimum formulations could be generated. The pros and cons of NN in case of the immediate-release tablets were highlighted by Borquin and their group members [11,34]. Later, Plumb et al. used the data that had been generated by these authors to analyze different NN programs and classes of training algorithms, respectively. Most anticipated models were further analyzed from each NN, and the obtained results have shown that no significant differences were observed [35].

Shao and their co-authors have reported that the generated neuro-fuzzy results had the advantages, i.e., the rules were present in a clear format [36]. Similarly, Rocksloh and their group members also have used the NN for optimizing crushing strength and disintegration time of a plant extract tablet [37]. Do et al. showed the advantages of using both NN and GA in the formulation of antacid tablets [38]. Later, Sathe and Venitz found the better quality of NNs over that of statistical models by observing the dissolution of diltiazem tablets [39].

NNs have also been enforced for a novel formulation of rifampicin and isoniazid for the treatment of tuberculosis in children [21,40,41]. Apart from this, solid dispersion formulation of ketoprofen was also developed using the NN and neuro-fuzzy methods, respectively [42]. The study has also been extended by the addition of a microemulsion [43].

9.9 IN PRODUCT DEVELOPMENT

In the development of various products, the major issue is with a multicomponent nature, including formulation optimization and process variables, respectively. The most functional property of ANN is its capability to generalize and help in fixing several problems [44]. These models have advanced fitting and anticipating abilities in contrast to solid dosage forms in order to investigate the effect of various formulation and compression parameters, respectively. In case of microemulsion-based drug delivery development, ANN models are employed for analyzing the phase behavior of quaternary microemulsion comprising oil, water, and two surfactants, respectively. These models can also be beneficial in the identification of aerosol behavior and in designing the pulmonary drug delivery systems [45].

9.10 IN PREDICTIVE TOXICOLOGY

The description of potential toxicity of drug candidates requires lots of animals, well-developed bioassay methods, and advanced laboratory facilities [46]. The complication of determining the carcinogenicity and mutagenicity includes the three main perspectives, viz, physical simulations using molecular modeling methods, proficient systems competent towards reasoning, and data mining system [47–49]. On the phenomena of learning from raw data, various kinds of available databases such as distributed structure searchable toxicity (DSSTox) database network or vitic toxicity database can be beneficial in this area [50, 51]. These databases have the ability to store various in-house toxicological results that can be reanalyzed using different techniques. These kinds of large data sets are necessary to develop and evaluate novel approaches due to their accuracy level around 80%–95%, which is almost comparable to in vivo assessments [52–53].

9.11 PROTEINS' FUNCTION AND STRUCTURE PREDICTION

The computational analysis based on the NN approach is good for analyzing molecular sequencing of data as well as its classification along with the identification of gene and the prediction of protein structure. For the further study of the relationship between the structure and the function of protein, the results are variable and the process is very convenient [54–55].

Livingstone et al. have described the various pros of networks in the simulation of drug molecules and protein structures along with a backpropagation that was usually employed for analyzing pattern recognition in the protein side-chain–side-chain contact maps [56,57]. Earlier several networks were formed to a set of models for the side chains of protein structures. These models were able to generate the results up to the accuracy level of 84.5%, and for the test suite, a Matthews's coefficient of around 0.72, respectively. ANNs in combination with GA training algorithms have been used for the determination of the secondary and foldable structures of RNA chains followed by their alignment and assembling along with the identification of the quantitative similarity between sequences, respectively [58–61].

9.12 PHARMACOKINETICS

Doses and drug options are resolute by understandings of concepts related to PK and pharmacodynamics (PD) of the drug. Often there is an insufficient information about the PK effect of a drug or drug's behavior for each individual patient. The ANN represents the new model-independent approach for the analysis of PK-PD data. These approaches can easily predict PD profiles without gathering any information regarding metabolites due to their independence of structural details, and also exhibit a certain level of advantage over other conventional methods [62–63].

According to one survey report, almost around 75 companies of the United Kingdom are holding business in this neural computing applications sector. It was also published that around 82%–85% users are very much satisfied with their systems and business benefits especially for the domain of product formulations due to an improvement in product quality at lower cost and shorter duration with the fast development of new formulations [64–65].

9.13 CONCLUSION

The applications of AI in pharmaceuticals are very vast, viz, pattern recognition and modeling analytical data, modeling the response surface, in formulation, in product development, in PK, and many more. There is a great deal of interest in neural computing, but the quantified information on the benefits has been harder to find. Here, we have discussed the role of AI or ANN in many aspects of pharmaceuticals. There are really so many interesting concepts that have been part of neural computing such as a rapid data analysis process, an ability to save and analyze the large data, an effective examination of the total design space irrespective of complexity followed by their applicability towards incomplete data sets with a better refinement along with a capacity to hold preferences and constraints as well to generate understandable rules.

REFERENCES

1. Chen, Y., W.T. McCall, R.A. Baichwal, C.M. Meyer, The application of an artificial neural network and pharmacokinetic simulations in the design of controlled-release dosage forms. *Journal of Controlled Release*, 1999. **59**(1): pp. 33–41.
2. Mahajan, A., et al., Artificial intelligence in healthcare in developing nations: The beginning of a transformative journey. *Cancer Research, Statistics, and Treatment*, 2019. **2**(2): p. 182.

3. Prital Sable, V.V.K., Pharmaceutical application of artificial intelligence. 2018: p. 2343.
4. Duch, W., K. Swaminathan, and J. Meller, Artificial intelligence approaches for rational drug design and discovery. *Current Pharmaceutical Design*, 2007. **13**(14): pp. 1497–1508.
5. Belič, A., et al., Minimisation of the capping tendency by tableting process optimisation with the application of artificial neural networks and fuzzy models. *European Journal of Pharmaceutics and Biopharmaceutics*, 2009. **73**(1): pp. 172–178.
6. Sutariya, V., et al., Artificial neural network in drug delivery and pharmaceutical research. *The Open Bioinformatics Journal*, 2013. **7**(1): pp. 49–62.
7. Chen, Y.-Y., et al., Design and implementation of cloud analytics-assisted smart power meters considering advanced artificial intelligence as edge analytics in demand-side management for smart homes. *Sensors*, 2019. **19**(9): p. 2047.
8. Gatys, L.A., A.S. Ecker, and M. Bethge, A neural algorithm of artistic style. arXiv preprint arXiv:1508.06576, 2015.
9. Agatonovic-Kustrin, S. and R. Beresford, Basic concepts of artificial neural network (ANN) modeling and its application in pharmaceutical research. *Journal of Pharmaceutical and Biomedical Analysis*, 2000. **22**(5): pp. 717–727.
10. Rowe, R. and E. Colbourn, Generating rules for tablet formulation. *Pharmaceutical Technology International*, 2000. **12**: pp. 24–27.
11. Bourquin, J., et al., Application of artificial neural networks (ANN) in the development of solid dosage forms. *Pharmaceutical Development and Technology*, 1997. **2**(2): pp. 111–121.
12. Sun, Y., et al., Application of artificial neural networks in the design of controlled release drug delivery systems. *Advanced Drug Delivery Reviews*, 2003. **55**(9): pp. 1201–1215.
13. Rowe, R.C. and R.J. Roberts, Artificial intelligence in pharmaceutical product formulation: neural computing and emerging technologies. *Pharmaceutical Science & Technology Today*, 1998. **1**(5): pp. 200–205.
14. Yadav, G., Y. Kumar, and G. Sahoo, Role of the computational intelligence in drugs discovery and design: introduction, techniques and software. *International Journal of Computer Applications*, 2012. **51**(10): pp. 7–18.
15. Bhargavi, P. and S. Jyothi, Soil classification using data mining techniques: a comparative study. International Journal of Engineering Trends and Technology, 2011. **2**: p. 5558.
16. Patel, V.L., et al., The coming of age of artificial intelligence in medicine. *Artificial Intelligence in Medicine*, 2009. **46**(1): pp. 5–17.
17. Aksu, B. and B. Mesut, Quality by design (QbD) for pharmaceutical area. *İstanbul Üniversitesi Eczacılık Fakültesi Dergisi*, 2015. **45**(2): pp. 233–251.
18. Hayes, C. and T. Gedeon, Hyperbolicity of the fixed point set for the simple genetic algorithm. *Theoretical Computer Science*, 2010. **411**(25): pp. 2368–2383.
19. Arabgol, S. and H.S. Ko, Application of artificial neural network and genetic algorithm to healthcarewaste prediction. *Journal of Artificial Intelligence and Soft Computing Research*, 2013. **3**(4): pp. 243–250.
20. Chakraborty, R., Fundamentals of genetic algorithms. *Reproduction*, 2010. **22**: p. 35.
21. Ibrić, S., et al., Artificial intelligence in pharmaceutical product formulation: neural computing. *CICEQ-Chemical Industry and Chemical Engineering Quarterly*, 2009. **15**(4): pp. 227–236.
22. VerDuin, W.H., *Better Products Faster: Practical Uses for Knowledge Based Systems in Manufacturing*. 1994: McGraw-Hill, Inc., London.
23. Walters, W. and B. Goldman, Feature selection in quantitative structure-activity relationships. *Current Opinion in Drug Discovery & Development*, 2005. **8**(3): pp. 329–333.
24. Bourquin, J., et al., Basic concepts of artificial neural networks (ANN) modeling in the application to pharmaceutical development. *Pharmaceutical Development and Technology*, 1997. **2**(2): pp. 95–109.

25. Agatonovic-Kustrin, S., et al., Application of neural networks for response surface modeling in HPLC optimization. *Analytica Chimica Acta*, 1998. **364**(1–3): pp. 265–273.
26. Mitchell, M., *An Introduction to Genetic Algorithms*. 1998: MIT Press, Cambridge.
27. Prital Sable, V.V.K., Pharmaceutical application of artificial intelligence. 2018: p. 2344.
29. Takayama, K., et al., Formula optimization of theophylline controlled-release tablet based on artificial neural networks. *Journal of Controlled Release*, 2000. **68**(2): pp. 175–186.
28. Bozič, D.Z., F. Vrečer, and F. Kozjek, Optimization of diclofenac sodium dissolution from sustained release formulations using an artificial neural network. *European Journal of Pharmaceutical Sciences*, 1997. **5**(3): p. 163–169.
30. Ibrić, S., et al., The application of generalized regression neural network in the modeling and optimization of aspirin extended release tablets with Eudragit® RS PO as matrix substance. *Journal of Controlled Release*, 2002. **82**(2–3): pp. 213–222.
31. Ibric, S., et al., Generalized regression neural networks in prediction of drug stability. *Journal of Pharmacy and Pharmacology*, 2007. **59**(5): pp. 745–750.
32. Türkoglu, M., R. Özarslan, and A. Sakr, Artificial neural network analysis of a direct compression tabletting study. *European Journal of Pharmaceutics and Biopharmaceutics*, 1995. **41**(5): pp. 315–322.
33. Kesavan, J.G. and G.E. Peck, Pharmaceutical granulation and tablet formulation using neural networks. *Pharmaceutical Development and Technology*, 1996. **1**(4): pp. 391–404.
34. Bourquin, J., et al., Advantages of Artificial Neural Networks (ANNs) as alternative modelling technique for data sets showing non-linear relationships using data from a galenical study on a solid dosage form. *European Journal of Pharmaceutical Sciences*, 1998. **7**(1): pp. 5–16.
35. Plumb, A.P., et al., Optimisation of the predictive ability of artificial neural network (ANN) models: a comparison of three ANN programs and four classes of training algorithm. *European Journal of Pharmaceutical Sciences*, 2005. **25**(4–5): pp. 395–405.
36. Shao, Q., R.C. Rowe, and P. York, Comparison of neurofuzzy logic and neural networks in modelling experimental data of an immediate release tablet formulation. *European Journal of Pharmaceutical Sciences*, 2006. **28**(5): p. 394–404.
37. Rocksloh, K., et al., Optimization of crushing strength and disintegration time of a high-dose plant extract tablet by neural networks. *Drug Development and Industrial Pharmacy*, 1999. **25**(9): pp. 1015–1025.
38. Do, Q., G. Dang, and N. Le, Le NQ. Drawing up and optimizing the formulation of Malumix tablets by an artificial intelligence system (CAD/Chem). *Tap Chi Duoc Hoc*, 2000. **6**: pp. 16–19.
39. Sathe, P.M. and J. Venitz, Comparison of neural network and multiple linear regression as dissolution predictors. *Drug Development and Industrial Pharmacy*, 2003. **29**(3): pp. 349–355.
40. Sunada, H. and Y. Bi, Preparation, evaluation and optimization of rapidly disintegrating tablets. *Powder Technology*, 2002. **122**(2–3): pp. 188–198.
41. Agatonovic-Kustrin, S., et al., Prediction of a stable microemulsion formulation for the oral delivery of a combination of antitubercular drugs using ANN methodology. *Pharmaceutical Research*, 2003. **20**(11): pp. 1760–1765.
42. Mendyk, A. and R. Jachowicz, Neural network as a decision support system in the development of pharmaceutical formulation – focus on solid dispersions. *Expert Systems with Applications*, 2005. **28**(2): pp. 285–294.
43. Mendyk, A. and R. Jachowicz, Unified methodology of neural analysis in decision support systems built for pharmaceutical technology. *Expert Systems with Applications*, 2007. **32**(4): pp. 1124–1131.
44. Man, K.-F., K.-S. Tang, and S. Kwong, *Genetic Algorithms: Concept and Design*. 1999: Springer Verlag, London.

45. Gen, M., R. Cheng, and L. Lin, *Network Models and Optimization: Multiobjective Genetic Algorithm Approach*. 2008: Springer Science & Business Media, London.
46. Kola, I. and J. Landis, Can the pharmaceutical industry reduce attrition rates? *Nature Reviews Drug discovery*, 2004. **3**(8): pp. 711–716.
47. Helma, C., *Predictive Toxicology*. 2005: CRC Press, Boca Raton, FL.
48. Cronin, M.T., *Predicting Chemical Toxicity and Fate*. 2004: CRC Press, Boca Raton, FL.
49. Benigni, R., *Quantitative Structure-Activity Relationship (QSAR) Models of Mutagens and Carcinogens*. 2003: CRC Press, Boca Raton, FL.
50. Hardy, B., Douglas, N., Helma, C., Rautenberg, M., Jeliazkova, N., et al., Collaborative development of predictive toxicology applications. Journal of cheminformatics. 2010 Dec. 2(1): pp. 1–29.
51. Lhasa Limited, http://www.lhasalimited.org assessed on 05.02.2020.
52. Matthews, E.J., et al., Assessment of the health effects of chemicals in humans: I. QSAR estimation of the maximum recommended therapeutic dose (MRTD) and no effect level (NOEL) of organic chemicals based on clinical trial data1. *Current Drug Discovery Technologies*, 2004. **1**(1): pp. 61–76.
53. Cheng, A. and S.L. Dixon, In silico models for the prediction of dose-dependent human hepatotoxicity. *Journal of Computer-Aided Molecular Design*, 2003. **17**(12): pp. 811–823.
54. Wu, C.H., Artificial neural networks for molecular sequence analysis. *Computers & Chemistry*, 1997. **21**(4): pp. 237–256.
55. Sun, Z., et al., Prediction of protein supersecondary structures based on the artificial neural network method. *Protein Engineering*, 1997. **10**(7): pp. 763–769.
56. Livingstone, D.J., D.T. Manallack, and I.V. Tetko, Data modelling with neural networks: advantages and limitations. *Journal of Computer-Aided Molecular Design*, 1997. **11**(2): pp. 135–142.
57. Milik, M., A. Kolinski, and J. Skolnick, Neural network system for the evaluation of side-chain packing in protein structures. *Protein Engineering, Design and Selection*, 1995. **8**(3): pp. 225–236.
58. Reidys, C., P.F. Stadler, and P. Schuster, Generic properties of combinatory maps: neutral networks of RNA secondary structures. *Bulletin of Mathematical Biology*, 1997. **59**(2): pp. 339–397.
59. Tacker, M., et al., Algorithm independent properties of RNA secondary structure predictions. *European Biophysics Journal*, 1996. **25**(2): pp. 115–130.
60. O'Neill, M.C., *A general procedure for locating and analyzing protein-binding sequence motifs in nucleic acids. Proceedings of the National Academy of Sciences*, 1998. **95**(18): pp. 10710–10715.
61. Sun, J., et al., Analysis of tRNA gene sequences by neural network. *Journal of Computational Biology*, 1995. **2**(3): pp. 409–416.
62. Gobburu, J.V. and E.P. Chen, Artificial neural networks as a novel approach to integrated pharmacokinetic – pharmacodynamic analysis. *Journal of Pharmaceutical Sciences*, 1996. **85**(5): pp. 505–510.
63. Turner, J.V., D.J. Maddalena, and D.J. Cutler, Pharmacokinetic parameter prediction from drug structure using artificial neural networks. *International Journal of Pharmaceutics*, 2004. **270**(1–2): pp. 209–219.
64. Rees, C., *Neural Computing –Learning Solutions – User Survey*. Department of Trade and Industry, London, 1996.
65. Rowe, R.C. and R.J. Roberts, *Intelligent Software for Product Formulation*. 1998: Taylor & Francis, Inc., London.

10 Role of Artificial Intelligence for Diagnosing Tuberculosis

Anshu Sharma
CT University

Anurag Sharma
GNA University

CONTENTS

10.1 INTRODUCTION

Tuberculosis (TB), which is perhaps the most established illness known to influence people and liable to have existed in pre-primates, is a significant reason for death around the world. This ailment is brought about by the microorganisms of the Mycobacterium tuberculosis (MTB) complex and normally influences the lungs; however, the different organs are also associated with it, which is found up to 33% cases [1]. Transmission for the most parts happens through the airborne spread of the tiny droplets delivered by patients with irresistible TB. TB is a sickness that has existed since relic. Worldwide, it is a significant danger to well-being, and it is the second most elevated reason for death from an irresistible malady after HIV/AIDS [2]. In economically developing countries, this malady is one of the chief reasons for casualties and sufferings [3]. The relatively high degree of control of TB in economically developed countries is due to the use of effective technology, but in economically developing countries, its control is becoming very challenging, and with the movement of populations, this problem could well spread to the economically developed countries and can result in a type of TB, which would be very difficult to control with the present methods of treatment [4]. Thus, the complete control of TB on a worldwide basis is therefore in everyone's interest.

10.1.1 History of TB

TB has probably been a human pathogen for millions of years. Skeletal remains are an important source of TB. The earliest evidence comes from a female skeleton aged 30 found in the cave of Arma Dell' Aquila in Liguria, Italy, dating back to around 5800 BC [5]. TB in mummified remains from Egypt was also noticed from 4500BC, with the most famous being that of a mummy Nesperhan in whom there is a clear evidence of both spinal changes and a psoas abscess.

Disease in Asia appears later with the earliest evidence being from 2700 BC. Skeletal proof of TB from the America is later going back to 1000 AD in North America and 700 AD in South America. The infectious nature of the disease was observed in 1546 when Girolamo Tracastoro described that bed sheets and wearables of a consumptive could have infectious particles. In 1865, the French specialist named Jean-Antoine Villemin proved that the TB was capable and could be spread from humans to animals. Though, in 1882, Robert Koch revealed that MTB was the cause of TB [5].

10.1.2 Global Impact of TB

TB is a fatal disease. World Health Organization (WHO), in 2018, presented statistics that stated around 10 million people suffered with TB, and about 1.5 million lost their lives due to this disease in 2018. In 2018, approximately 10 million people were affected with TB worldwide. Out of those 10 million individuals, 5.8 million were men, 3.5 million were ladies, and 1.0 million were kids. This gauge included cases all through the world and among all different age groups [6].

Evaluations of TB occurrence in 2018 disaggregated by age and sex are represented in Figure 10.1 (worldwide).People in all age bunches are influenced by TB yet the most noteworthy weight is among the grown-up men [7].

They represented 57% of all cases in 2018, in contrast with 32% of the cases in the grown-up ladies and 11% in kids [6]. Those reports available depicts that the cases of TB are higher among men [8]. The gravity of national TB is causing a great threat as number of TB cases relatively to the population size is increasing at an alarming rate among different countries in 2018. (Figure 10.2).

There were under ten event cases for every 100,000 masses in the most developed and wealthy countries;150–400 in most of the 30 high TB inconvenience countries; or increasingly 500 in the Central African Republic, the Democratic People's Republic of Korea, Lesotho, Mozambique, Namibia, the Philippines, and South Africa. Among the 30 high TB inconvenience countries, there were three with remarkably lower event rates per capita: China, Brazil, and the Russian Federation, which had best checks of 61, 45, and 54, independently [6].

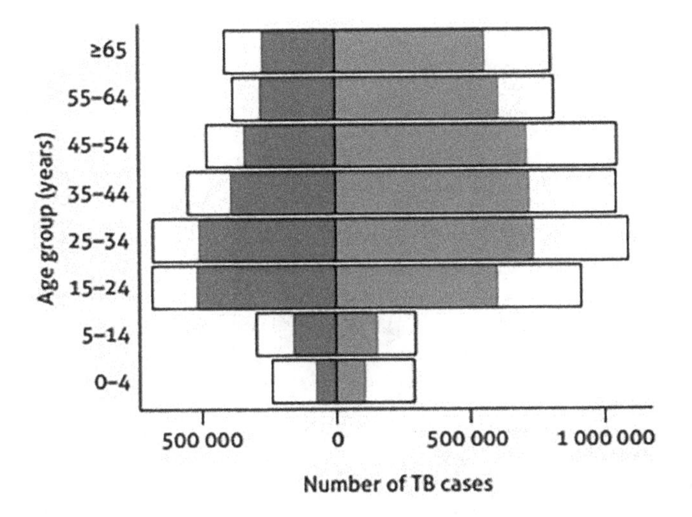

FIGURE 10.1 Global estimates of TB incidence (black outline) and case notifications disaggregated by age and sex (female in red; male in turquoise), 2018 [6].

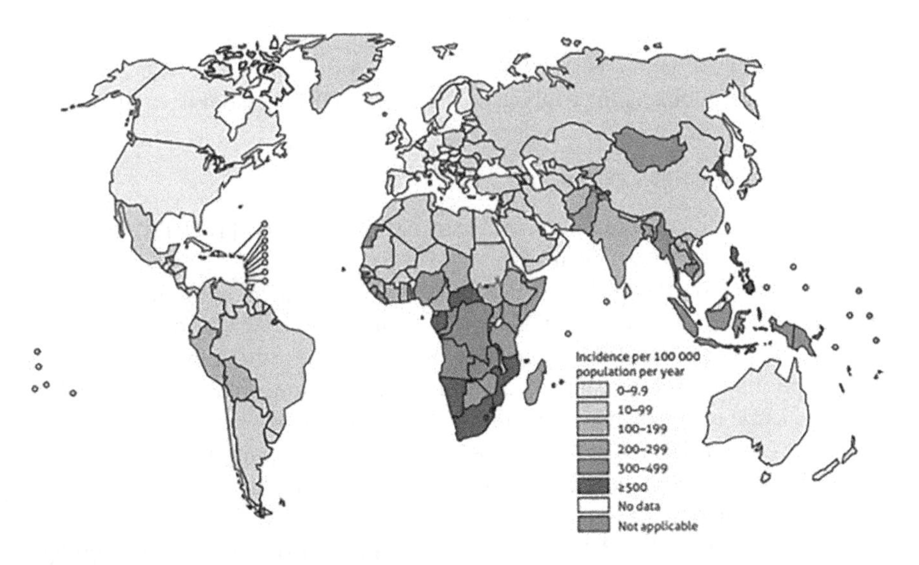

FIGURE 10.2 Estimated TB incidence rates in 2018 [6].

10.1.3 TB: India's Silent Epidemic

As of 2018, India accounts for the world's highest number of people suffering from the disease. In 2018, 21.5 lakh TB cases were found in India. 89% of the TB cases are found in the age group between 15 and69 years. Uttar Pradesh is the largest contributor of TB (20%). Two states, namely, Delhi and Chandigarh, stand aloof from all other states and UTs with regard to the notification rate [9].

Figure 10.3 represents the state-wise number of TB cases in percentage of population reported in 2018. It can be observed from the chart that Uttar Pradesh, with 17%

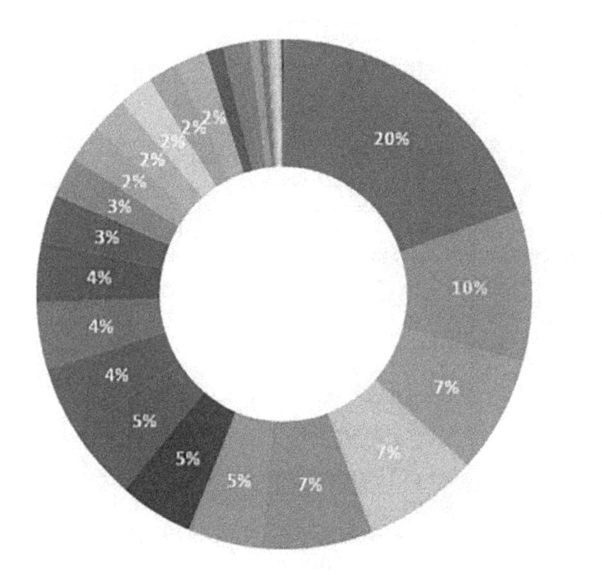

FIGURE 10.3 State-wise TB rate of India in 2018 [9].

of population of the nation, is the biggest supporter of the TB cases with 20% of the absolute notices, bookkeeping to about 4.2 lakh cases (187 cases/lakh population) [9].

10.1.4 Classification of TB

TB can be broadly classified into two categories: pulmonary TB (PTB) and extra-pulmonary TB (EPTB), which are explained as follows:

Pulmonary TB (PTB): PTB is characterized as a functioning disease of the lungs. It is the most significant TB disease, in light of the fact that a contamination of the lungs is profoundly infectious because of the method of bead transmission [10]. It very well may be life-threatening as well as perilous to the patient whenever left untreated. Little regions in the lung tainted with bacilli slowly form a structure filled up with the contaminated material [11]. A relentless cough, alongside significant side effects like perspiring in the night, fever, or inadvertent weight reduction, is the most well-known symptom of PTB [12] [13].

Extra-Pulmonary TB (EPTB): EPTB portrays the different conditions brought about by MTB contamination of organs or tissues outside the lungs. Arranged by recurrence, the EPTB destinations most ordinarily engaged with TB are the lymph hubs, pleura, bones, joints, and so on. Nevertheless, for all intents and purposes, all organ frameworks might be influenced [14].

10.2 TECHNOLOGICAL INTERVENTIONS FOR DIAGNOSIS OF TB

For the developing countries like India, there is a critical need to have moderate, convenient, and quick tests for TB conclusion. The ordinary strategies for the conclusion of TB are over 100 years of age that takes around 3–6 weeks to yield results [15].

For drug susceptibility tests, this may take even longer. This prompts a drawn-out postponement in analysis, at last bringing about a deferred treatment, which could worsen the course of the infection [16]. Moreover, directly observed therapy has been the typical approach to ensure patient's recovery throughout their treatment duration and analyzing adverse drug effects [17]. However, it results in a challenge for medical professionals to ensure the adherence of patients to the line of treatment as the patients need to visit healthcare workers fortnightly, which are sometimes not followed by patients properly.

Recently, with the quick advancement of data innovation and the developing regard for interdisciplinary practices, the utilization of computerized advances has become another territory of enthusiasm for clinical experts [18]. Moreover, in regard to the WHO's end TB blueprint and the advances in computerized technologies, there is a need to comprehend what's going on around the globe in regard to examination into the utilization of advanced innovation for better TB care and control [19].

Figure 10.4 depicts the contributions that can be offered by digital technology to accomplish the care and control of TB.

The digital technology can be broadly categorized to perform the following tasks:

a. **TB Diagnosis**: This can be achieved using artificial intelligence (AI) [20].
b. **TB Care and Control**: This can be achieved using IoT-based sensors, mobile application-based monitoring, and web-based monitoring.

Thus, it can be said that AI plays a vital role in the diagnosis of TB, and IoT plays a significant role in monitoring of TB patients. The brief explanation of these technologies has been explained below.

10.2.1 Artificial Intelligence (AI)

AI is defined as the simulation process using which machines are programmed in such a way that they think like humans and perform operations like human beings [21].

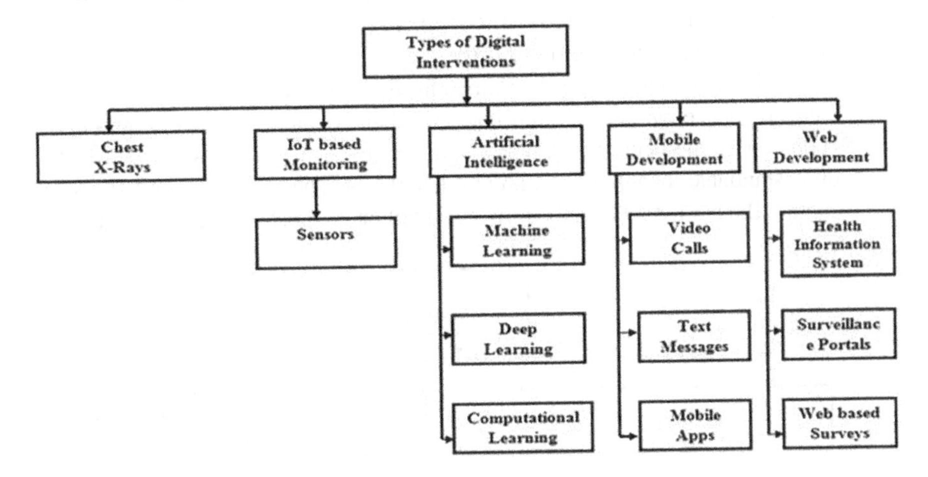

FIGURE 10.4 Types of digital interventions.

The term was previously shown in the Dartmouth Summer Research Project on AI in 1956 [21]. The central AI process accentuation is based on the three perspectives: picking up, thinking, and self-remedy.

a. **Learning Processes**: In this part of AI, the primary center is to obtain information and producing rules so as to change over the information into the significant structure [22].
b. **Reasoning Processes**: In this part of AI, the essential center is to pick the correct calculation to accomplish the ideal result.
c. Self-Correction Processes: In this part of AI, the essential center is to consistently refine and reclassify the calculations to guarantee that they give the exact outcomes as much as could be expected [23,24].

10.2.2 AI Techniques

AI can be categorized on the basis of machine's capacity to predict future decisions by using past experiences in order to create self-awareness and memory. Below are the various categories of AI (Figure 10.5):

a. Machine Learning

   Machine learning is one of the significant techniques of AI using machines that are made to learn and perform tasks without programming them explicitly. Deep learning, which is a part of machine learning, is used to predict data with the help of artificial neural networks [25]. Machine learning algorithms are categorized into unsupervised learning, supervised learning, and reinforcement learning. Unsupervised learning, as the name suggests, does not need any preclassification information to predict data [26]. Supervised learning predicts the data using the training function by employing machine that learns on the basis of input [27]. Reinforcement learning is a part of machine learning in which rigorous actions are taken continuously so as to get the best possible solution of the problem [28].
b. Natural Language Processing (NLP)

   In NLP, computers are programmed in such a way so that they can interact with human beings in a natural understandable form. The learning process is done with the help of machine learning in order to obtain the correct and understandable meaning from human languages. The first step in NLP is to capture the human talk (in the form of audio) into a machine. Then, this audio

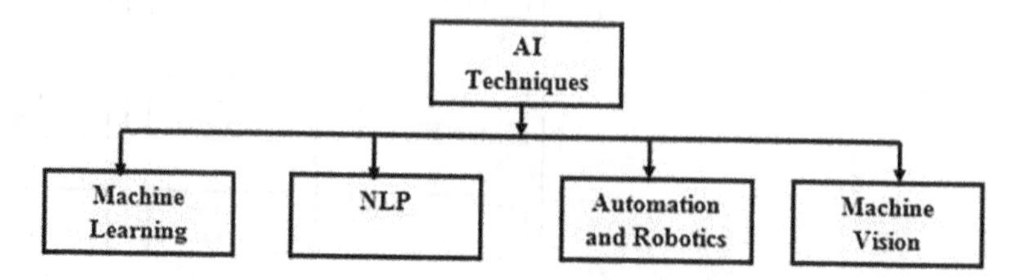

FIGURE 10.5 AI techniques.

is converted into the text, and then again, data is converted into audio [29]. After then, this converted audio is used by machines in the form of response to users (humans). Language translation applications like Google Translator and Grammar check facility of word processors are the suitable examples of NLP. However, the nature of human dialects makes the NLP troublesome as a result of the principles that are engaged to deal with the data for utilizing normal language, and they are difficult for the systems to comprehend [30]. So NLP utilizes the algorithms to perceive and extract the guidelines of the characteristic dialects where the unstructured information from the human dialects can be changed over to an organization that is comprehended by the computers.

c. Automation and Robotics

The motive behind automation is to get the tedious and tiresome assignments done by machines, which likewise improve profitability and result in getting financially savvy and progressively proficient outcomes. Numerous associations use the applications of machine learning, neural systems, and graphs for the computerization of their tasks [31]. Such mechanization can prevent deception issues while performing financial transactions online by using CAPTCHA development. Programming and automation are usually done to perform repetitive tasks, which can be different in different situations.

d. Machine Vision

Machines can catch visual data, and afterward, examination of this information is done. Here, cameras are utilized to catch the visual data; the simple-to-advanced transformation is utilized to change over the picture to computerized information, and computerized signal preparation is utilized to process the information. By then, the ensuing data is dealt with to a PC. In the machine vision, the two basic edges are affectability, which is the limit of the machine to see pixels that are not solid, and less goals, the range to which the machine can perceive the things [32,33]. The utilization of the machine vision can be found in signature unmistakable verification, plan affirmation, and clinical picture assessment etc.

10.2.3 Role of AI in the Diagnosis of TB—Comparative Analysis

Altogether to get the knowledge into the job of AI in the finding of TB and to get a diagram of the distributions covering this subject, a perusing survey has been done as such. The after-effects of this examination could at last be applied to upgrade the utilization of AI innovation in TB control all the more reasonably and viably [34]. To distinguish every single important investigation, a far-reaching search procedure was created to recover the writing identified with TB and its determination utilizing AI. These hunt terms were utilized to recognize applicable writing in two essential databases, Web of Science and PubMed.

Table 10.1 represents the comparative analysis of various AI techniques that have been used to diagnose TB. It has been observed from Table10.1 that different AI techniques can be used to diagnose TB. Most commonly used techniques involve convolutional neural networks, Random Forest classifier, fuzzy inference system, artificial neural networks etc.

TABLE 10.1
Diagnosis of TB Using AI-Comparative Study

S. No	Paper Title	Journal/Conference	Author (Year)	AI Technique Used	Findings
1	Tuberculosis (TB) location framework utilizing profound neural systems	*Neural Computing and Applications*	Samuel et al. (2018)	Inception V3 and support vector machine	Two subsystems are proposed for the identification of TB, for example, data acquisition system and recognition system. This proposed framework pre-prepared loads of Inception V3 and utilized the help vector machine for characterizing the information. This model had an accuracy of 95.05% [35]
2	Early detection of TB using chest X-ray (CXR) with computer-aided diagnosis	*IEEE*	Gabriella et al. (2018)	Naïve Bayes classification	Developed a computer-aided diagnosis (CADx) to help specialists and radiologists for the early analysis of TB. In view of the calculations, the systems' sensitivity was determined as 66.67%, with specificity at 86% and accuracy at 76% [36]
3	Towards enlightening model for TB lesion discrimination on X-ray CT images	*IEEE 15th International Symposium on Biomedical Imaging*	Gordaliza et. al. (2018)	Random Forest classifier	Proposed the technique to distinguish TB lesions utilizing statistical region merging. The creators came to the resolution that the proposed framework is utilized for choosing simpler and less difficult yet predominant model that helps for the arrangement of strange tissue all the more precisely [37]
4	Towards automated tuberculosis detection using deep learning	*Computational Intelligence*	Kant et al. (2018)	Deep neural networks	Proposed a procedure that takes zoomed microscopic picture of sputum as info and returns the situation of dubious Mycobacterium TB bacilli as a result. This proposed framework with the assistance of microscopic picture of sputum for the recognition of TB accomplished 83.78% and 67.55% of review and accuracy separately [38]

(*Continued*)

TABLE 10.1 (*Continued*)
Diagnosis of TB Using AI-Comparative Study

S. No	Paper Title	Journal/Conference	Author (Year)	AI Technique Used	Findings
5	Automatic detection of tuberculosis bacilli from microscopic sputum smear images using the deep learning methods	*International Journal on Biocybernetics and Biomedical Engineering*	Panicker et al. (2018)	Convolutional neural network	The proposed strategy performs location of TB, by picture binarization and the resulting arrangement of recognized regions utilizing a convolutional neural system. Test results reveals that the proposed calculation accomplishes 97.13% review, 78.4% exactness and 86.76% F-score for the TB recognition. The proposed technique consequently distinguishes whether the sputum smear picture is contaminated with TB or not [39]
6	Automatic detection of mycobacterium TB using artificial intelligence	*Journal of Thoracic Disease*	Xiong et al. (2018)	Convolutional neural networks	Built a convolutional neural network (CNN) model, named TB AI (TB-AI), explicitly to perceive TB bacillus. Prepared TB-AI was run on the test information twice. TB-AI accomplished 97.94% sensitivity and 83.65% specificity [40]
7	Development of a rapid and sensitive DNA turbidity biosensor test for the diagnosis of katG gene in isoniazid-resistant Mycobacterium tuberculosis	*IEEE Sensors*	Ckumdee et al. (2017)	Novel loop-mediated isothermal amplification	Proposed a framework that utilized the calculation named as novel loop-intervened isothermal enhancement, which distinguishes one base point, which prompts the change of opposition of medication in Mycobacterium TB. The examples from 30 clinical were additionally tried, and the results demonstrated 100% sensitivity and specificity [41]

(*Continued*)

TABLE 10.1 (*Continued*)
Diagnosis of TB Using AI-Comparative Study

S. No	Paper Title	Journal/Conference	Author (Year)	AI Technique Used	Findings
8	Mycobacterium TB modeling using regression analysis	*IEEE Symposium on Computer Applications & Industrial Electronics (ISCAIE)*	Radzi et al. (2017)	Regression analysis	Projected a nonintrusive electronic-based identification instrument, which utilizes the regression analysis that delineates about the displaying of the Mycobacterium TB of delicate kind. The least blunder and variation given by the second-order LC circuit is least [42]
9	An automated TB screening strategy combining X-ray-based computer-aided detection and clinical information	*Scientific Reports*	Melendez et al. (2016)	Multiple learner fusion	Designed a component that utilized the one-class arrangement and decreased the gap with the administered approach. The proposed blend system beats the individual techniques in wording specificity at 95% [43]
10	Automatic detection of pulmonary TB using image processing techniques	*International Conference on Wireless Communications, Signal Processing and Networking*	Poornimadevi et al. (2016)	Watershed algorithm using artificial neural networks	Presented a computerized approach utilizing chest radiograph pictures to identify TB. The exactness of the proposed technique is 60% contrasted with active contour and worldwide thresholding [44]
11	Microscopic image segmentation based on the firefly algorithm for the detection of TB bacteria	*Signal Processing and Communications Applications Conference*	Ayas et al. (2015)	Firefly algorithm based on Swarm intelligence	Firefly algorithm IS proposed is dependent on Swarm knowledge to fragment pictures. Division results are contrasted and expert-guided division results and accomplished the presentation proportion of 96% [45]
12	Design and identification of TB using fuzzy-based decision support system	*Advances in Computer Science and Information Technology*	Walia et al. (2015)	Fuzzy inference system	Conducted an investigation based on soft computing methods, including artificial neural network and fuzzy interference systems that have been widely acknowledged to be accepted to model expert behavior [46]

(*Continued*)

TABLE 10.1 (*Continued*)
Diagnosis of TB Using AI-Comparative Study

S. No	Paper Title	Journal/Conference	Author (Year)	AI Technique Used	Findings
13	A decision support system for TB diagnosability	*International Journal on Soft Computing*	Walia et al. (2015)	Decision tree with fuzzy expert system	Proposed a fuzzy diagnosability approach, which takes an incentive between {0, 1} and dependent on perceptibility of occasions, the authors formalized the development of findings that are utilized to perform the determination. Exactness investigation of designed decision support system supportive network dependent on the segment information was finished by contrasting expert knowledge and system-generated response [47]
14	A fuzzy expert system for the prevention and diagnosis of blood diseases	*Indian Journal of Fundamental and Applied Life Sciences*	Katebi et al. (2014)	Machine learning	Examined the significance of AI to look through the presence of Cocci microscopic organisms. The most critical strategies utilized for characterizing the microscopic organisms depend on bacterium shape and cell arrangement. The proposed strategies can give best outcomes and would fundamentally help for grouping of microscopic organisms [48]
15	A survey on the applications of fuzzy logic in medical diagnosis	*International Journal of Scientific & Engineering Research*	Prasath et al. (2013)	Fuzzy expert system	Surveyed on numerous clinical decision support system constructions with the proper use of artificial intelligence. Remembering the investigation, by the use of fuzzy control, future improvements were prescribed to screen the exercises for the forecasting of the ailment effectively [49]
16	A soft computing genetic-neuro-fuzzy approach for data mining and its application to medical diagnosis	*International Journal of Engineering Advancement and Technology*	Rawat et al. (2013)	Neuro-fuzzy inference system	Suggested that a joined hereditary versatile neuro-fuzzy interference approach for information mining and its application to clinical analysis was investigated. The proposed strategies were figured on ovarian malignant growth dataset to amplify precision and limit cost. The proposed technique gives quick execution results [50]

(*Continued*)

TABLE 10.1 (*Continued*)
Diagnosis of TB Using AI-Comparative Study

S. No	Paper Title	Journal/Conference	Author (Year)	AI Technique Used	Findings
17	TB disease diagnosis by using adaptive neuro-fuzzy inference system and rough sets	*Neural Computing and Applications*	Ucar et al. (2013)	Neuro-fuzzy inference system	Evaluated the presentation of sugeno-type versatile neuro-fuzzy interface framework and rough set technique to anticipate the presence of mycobacterium TB on presumed patients. The reproduction strengthening shows that creating versatile framework characterizes the patients with an accuracy of 97% [51]
18	A rule-based fuzzy diagnostics decision support system for tuberculosis	*International Conference on Software Engineering Research, Management and Applications*	Soundaranjan et al. (2011)	Rule-based fuzzy expert system	Explained that a rule-based fuzzy diagnostics emotionally supportive network for TB detection. The proposed framework engineering and calculation decides the likely class of TB utilizing rule-based methodology [52]
19	Automated TB screening using image processing tools	*Pan American Health Care Exchanges*	Castaneda et al. (2010)	Random Forest algorithm	Utilized the image processing algorithms to distinguish and analyze TB in tiny pictures of sputum tests recolored with the Ziehl-Neelsen strategy [53]
20	Support vector machines for automatic detection of tuberculosis bacteria in confocal microscopy images	*2007 4th IEEE International Symposium on Biomedical Imaging: From Nano to Macro*	Lenseigne et al. (2007)	Support vector machine	The authors played out an image segmentation dependent on support vector machines so as to group a picture at a pixel level. At the same time, the authors also applied this technique to measure the amount of Mycobacterium tuberculosis in confocal microscopy pictures for drug discovery inside the setting of high content screening (HCS). To manage the presentation limitations of HCS, the authors suggested a model selection algorithm that finds the best classifier's hyperparameters by enhancing both classification rate and complexity [54]

TABLE 10.2
Comparative Analysis on the Basis of Accuracy

S. No	AI Techniques Used	Maximum Accuracy Achieved (%)
1	Convolutional neural networks	86
2	Watershed algorithm using artificial neural networks	60
3	Fuzzy inference system	94
4	Random Forest classifier	92
5	Support vector machine	95
6	Naïve Bayes algorithm	76
7	Deep neural networks	83
8	Multiple learner fusion	95
9	Rule-based fuzzy system	97
10	Firefly algorithm based on swarm intelligence	96

Table 10.2 represents the comparative analysis of the commonly used AI techniques and their accuracy in diagnosing TB. But accuracy cannot be considered as the only parameter to determine the efficiency of the system as it largely depends on the number of data items included. It has been observed that in some cases that the system performs more accurately for a less number of data items.

10.2.4 Limitations of Retrieved Literature

The above literature has some limitations. Potential research subjects have not been very much examined till date, which include the diagnosis of EPTB that has not been investigated by researchers to a large extent.

Furthermore, surveillance of TB diagnosis and TB drug forecasting are the areas that need to be explored as these play a vital role in controlling the growth of TB.

Moreover, only few apps were developed for use by patients, and none were designed to support TB patients' involvement in and management of their care.

Therefore, it can be said that there is a need to investigate using AI techniques that can be used to diagnose both PTB and EPTB. Also, a monitoring system can be designed using IoT that can be used by patients and doctors to monitor the recovery of patients.

10.3 CONCLUSION

No doubt AI proves a promising supportive network to diagnose TB and can help clinicians to make decisions regarding the diagnosis of TB. Also, it holds the possibility to mitigate the overwhelming remaining burden of pathologists and lessening odds of missed finding. However, still there is a need for the research in the diagnosis of EPTB, which has not been covered in most of the literature. Patient care and monitoring is one of the most important tasks for the clinicians, which needs to be explored and researched in order to make an easier communication between patients and doctors.

REFERENCES

1. F. Rageade, N. Picot, A. Blanc-Michaud, S. Chatellier, C. Mirande, E. Fortin, and A. van Belkum, "Performance of solid and liquid culture media for the detection of *Mycobacterium tuberculosis* in clinical materials: Meta-analysis of recent studies," *European Journal of Clinical Microbiology & Infectious Diseases*, vol. 33, pp. 867–870, 2014.
2. B. M. Mayosi, M. Ntsekhe, J. Bosch, S. Pandie, H. Jung, F. Gumedze, J. Pogue, L. Thabane, M. Smieja, V. Francis, and L. Joldersma, "Prednisolone and *Mycobacterium indicus pranii* in *Tuberculous pericarditis*," *The New England Journal of Medicine*, vol. 371, no. 12, pp. 1121–1130, 2014.
3. A. Frankel, C. Penrose, and J. Emer, "Cutaneous tuberculosis: A practical case report and review for the dermatologist," *The Journal of Clinical and Aesthetic Dermatology*, vol. 2, no. 10, pp. 19–27, 2009.
4. A. Polaski, S. E. Tatro, and J. Luckmann, *Luckmann's Core Principles and Practice of Medical-Surgical Nursing*. Saunders, Philadelphia, PA, 1996.
5. J. H. Grosset and R. E. Chaisson, eds., *Handbook of Tuberculosis*. Springer International Publishing, Cham, 2017.
6. World Health Organization, "10 facts on tuberculosis," 2019. [Online].https://www.who.int/features/factfiles/tuberculosis/en/.
7. Public Health Concern, "National tuberculosis management guidelines 2019," *Public Health Concern (PHC)*, 27-Oct-2019. [Online]. Available: https://publichealthconcern.com/national-tuberculosis-management-guidelines-2019/. [Accessed: 17-May-2020].
8. "Health Library, CS Mott Children's Hospital, Michigan Medicine,"*Mottchildren.Org*, 2019. https://www.mottchildren.org/health-library/hw207130.
9. M. India, "Annual reports: Central TB Division," *Tbcindia.gov.in*, 2019. [Online]. Available at: https://tbcindia.gov.in/index1.php?lang=1&level=1&sublinkid=4160&lid=2807.
10. L. Ji, Y.-L. Lou, Z.-X. Wu, J.-Q. Jiang, X.-L. Fan, L.-F. Wang, X.-X. Liu, P. Du, J. Yan, and A.-H. Sun, "Usefulness of interferon-γ release assay for the diagnosis of sputum smear-negative pulmonary and extra-pulmonary TB in Zhejiang Province, China," *Infectious Diseases of Poverty*, vol. 6, no. 1, pp. 1–5, 2017.
11. N. A. Mohamad, N. A. Jusoh, Z. Z. Htike, and S. L. Win, "Bacteria identification from microscopic morphology: A survey," *International Journal on Soft Computing, Artificial Intelligence and Applications*, vol. 3, no. 2, pp. 1–12, 2014.
12. Cmt and Tac, "Pulmonary TB," *TB Online – Pulmonary TB*. [Online]. Available at: http://www.tbonline.info/posts/2016/3/31/pulmonary-tb/. [Accessed: 17-May-2020].
13. G. García-Elorriaga and G. del Rey-Pineda, *Practical and Laboratory Diagnosis of Tuberculosis: From Sputum Smear to Molecular Biology*. Springer International Publishing, Cham, 2015.
14. S. K. Sharma, H. Ryan, S. Khaparde, K. S. Sachdeva, A. D. Singh, A. Mohan, R. Sarin, C. N. Paramasivan, P. Kumar, N. Nischal, S. Khatiwada, P. Garner, and P. Tharyan, "Index-TB guidelines: Guidelines on extrapulmonary tuberculosis for India," *The Indian Journal of Medical Research*, 2017. [Online]. Available: https://www.ncbi.nlm.nih.gov/pmc/articles/PMC5663158/. [Accessed: 17-May-2020].
15. Y. Payasi and S. Patidar, "Diagnosis and counting of tuberculosis bacilli using digital image processing," *2017 International Conference on Information, Communication, Instrumentation and Control (ICICIC)*, 2017. IEEE, Indore, India.
16. R. Zheng, C. Zhu, Q. Guo, L. Qin, J. Wang, J. Lu, H. Cui, Z. Cui, B. Ge, J. Liu, and Z. Hu, "Pyrosequencing for rapid detection of tuberculosis resistance in clinical isolates and sputum samples from re-treatment pulmonary tuberculosis patients," *BMC Infectious Diseases*, vol. 14, no. 1, pp. 200, 2014.

17. T. Uçar, A. Karahoca, and D. Karahoca, "Tuberculosis disease diagnosis by using adaptive neuro fuzzy inference system and rough sets," *Neural Computing and Applications*, vol. 23, no. 2, pp. 471–483, 2012.
18. M. S. Hossain, A. A. Monrat, M. Hasan, R. Karim, T. A. Bhuiyan, and M. S. Khalid, "A belief rule-based expert system to assess mental disorder under uncertainty," *2016 5th International Conference on Informatics, Electronics and Vision (ICIEV)*, 2016.IEEE, Dhaka, Bangladesh.
19. S. Jaeger, A. Karargyris, S. Candemir, L. Folio, J. Siegelman, F. Callaghan, Z. Xue, K. Palaniappan, R. K. Singh, S. Antani, and G. Thoma, "Automatic tuberculosis screening using chest radiographs," *IEEE Transactions on Medical Imaging*, vol. 33, no. 2, pp. 233–245, 2013.
20. C. Wattal and R. Raveendran, "Newer diagnostic tests and their application in pediatric TB," *The Indian Journal of Pediatrics*, vol. 86, no. 5, pp. 441–447, 2019.
21. R. Solomonoff, "The time scale of artificial intelligence: Reflections on social effects," *Human Systems Management*, vol. 5, no. 2, pp. 149–153, 1985.
22. R. Khutlang, S. Krishnan, A. Whitelaw, and T.S. Douglas, "Detection of tuberculosis in sputum smear images using two one-class classifiers,"*2009 IEEE International Symposium on Biomedical Imaging: From Nano to Macro, 2009.* IEEE, Boston, MA.
23. A. Karargyris, S. Antani, and G. Thoma, "Segmenting anatomy in chest x-rays for tuberculosis screening,"*2011 Annual International Conference of the IEEE Engineering in Medicine and Biology Society*, 2011. IEEE, Boston, MA.
24. N.M. Noor, O.M. Rijal, A. Yunus, A.A. Mahayiddin, G.C. Peng, and S.A.R. Abu-Bakar, "A statistical interpretation of the chest radiograph for the detection of pulmonary tuberculosis,"*2010 IEEE EMBS Conference on Biomedical Engineering and Sciences (IECBES)*, 2010. IEEE, Kuala Lumpur, Malaysia.
25. V. Ayma, R. De Lamare, and B. Castañeda, "An adaptive filtering approach for segmentation of tuberculosis bacteria in Ziehl-Neelsen sputum stained images,"*2015 Latin America Congress on Computational Intelligence (LA-CCI)*, 2015. IEEE, Curitiba, Brazil.
26. M. A. Saeed, S. M. Khan, N. Ahmed, M. U. Khan, and A. Rehman, "Design and analysis of capacitance based Bio-MEMS cantilever sensor for tuberculosis detection," *2016 International Conference on Intelligent Systems Engineering (ICISE)*, 2016. IEEE, Islamabad, Pakistan.
27. T. Ahmed, Md. Ferdous Wahid, and Md. Jahid Hasan, "Combining deep convolutional neural network with support vector machine to classify microscopic bacteria images,"*2019 International Conference on Electrical, Computer and Communication Engineering (ECCE)*, 2019. IEEE, Cox's Bazar, Bangladesh.
28. A. R. C. Semogan, B. D. Gerardo, B. T. Tanguilig III, J. T. D. Castro, and L. F. Cervantes, "A rule-based fuzzy diagnostics decision support system for tuberculosis," *2011 Ninth International Conference on Software Engineering Research, Management and Applications*, 2011. IEEE, Baltimore, MD.
29. Ch. Schuh, "Fuzzy sets and their application in medicine," *NAFIPS 2005–2005 Annual Meeting of the North American Fuzzy Information Processing Society*, 2005. IEEE, Detroit, MI.
30. T. Kaewphinit, N. Arunrut, W. Kiatpathomchai, S. Santiwatanakul, P. Jaratsing, and K. Chansiri, "Detection of *Mycobacterium tuberculosis* by using loop-mediated isothermal amplification combined with a lateral flow dipstick in clinical samples," *BioMed Research International*, vol. 2013, p. 926230, 2013.
31. H. El-Samadony, H.M. Azzazy, M.A. Tageldin, M.E. Ashour, I.M. Deraz, and T. Elmaghraby, "Nanogold assay improves accuracy of conventional TB diagnostics," *Lung*, vol. 197, no.2, pp.241–247, 2019.

32. R. S. Chithra and P. Jagatheeswari, "Fractional crow search-based support vector neural network for patient classification and severity analysis of tuberculosis," *IET Image Processing*, vol. 13, no. 1, pp. 108–117, 2018.
33. Y. P. López, C. F. F. Costa Filho, L. M. R. Aguilera, and M. G. F. Costa, "Automatic classification of light field smear microscopy patches using convolutional neural networks for identifying *Mycobacterium tuberculosis*,"*2017 CHILEAN Conference on Electrical, Electronics Engineering, Information and Communication Technologies (CHILECON)*. IEEE, 2017. Pucón, Chile.
34. A. J. Jara, M. A. Zamora, and A. F. Skarmeta, "Drug identification and interaction checker based on IoT to minimize adverse drug reactions and improve drug compliance," *Personal and Ubiquitous Computing*, vol. 18, no. 1, pp. 5–17, 2012.
35. R. D. J. Samuel and B. R. Kanna, "Tuberculosis (TB) detection system using deep neural networks," *Neural Computing and Applications*, vol. 31, no. 5, pp. 1533–1545, 2018.
36. I. Gabriella, S. A. Kamarga, and A. W. Setiawan, "Early detection of tuberculosis using chest X-Ray (CXR) with computer-aided diagnosis," *2018 2nd International Conference on Biomedical Engineering (IBIOMED)*, 2018. IEEE, Kuta, Indonesia.
37. P. M. Gordaliza, J. Vaquero, S. Sharpe, M. Desco, and A. Munoz-Barrutia, "Towards an informational model for tuberculosis lesion discrimination on X-ray CT images," *2018 IEEE 15th International Symposium on Biomedical Imaging (ISBI 2018)*, 2018. IEEE, Washington, DC.
38. S. Kant and M. M. Srivastava, "Towards automated tuberculosis detection using deep learning," *2018 IEEE Symposium Series on Computational Intelligence (SSCI)*, 2018. IEEE, Bangalore, India.
39. R. O. Panicker, K. S. Kalmady, J. Rajan, and M. Sabu, "Automatic detection of tuberculosis bacilli from microscopic sputum smear images using deep learning methods," *Biocybernetics and Biomedical Engineering*, vol. 38, no. 3, pp. 691–699, 2018.
40. Y. Xiong, X. Ba, A. Hou, K. Zhang, L. Chen, and T. Li, "Automatic detection of *Mycobacterium tuberculosis* using artificial intelligence," *Journal of Thoracic Disease*, vol. 10, no. 3, pp. 1936–1940, 2018.
41. J. Ckumdee, S. Santiwatanakul, and T. Kaewphinit, "Development of a rapid and sensitive DNA turbidity biosensor test for diagnosis of katG gene in isoniazid resistant *Mycobacterium tuberculosis*," *2017 IEEE Sensors*, 2017. Glasgow, UK.
42. R. R. M. Radzi, W. Mansor, and J. Johari, "Mycobacterium tuberculosis modelling using regression analysis," *2017 IEEE Symposium on Computer Applications & Industrial Electronics (ISCAIE)*, 2017. IEEE, Langkawi, Malaysia.
43. J. Melendez, C. I. Sánchez, R. H. H. M. Philipsen, P. Maduskar, R. Dawson, G. Theron, K. Dheda, and B. V. Ginneken, "An automated tuberculosis screening strategy combining X-ray-based computer-aided detection and clinical information," *Scientific Reports*, vol. 6, no. 1, pp. 1–8, 2016.
44. C. S. Poornimadevi and C. Helen Sulochana, "Automatic detection of pulmonary tuberculosis using image processing techniques," *2016 International Conference on Wireless*
45. S. Ayas, H. Dogan, E. Gedikli, and M. Ekinci, "Microscopic image segmentation based on firefly algorithm for detection of tuberculosis bacteria," *2015 23nd Signal Processing and Communications Applications Conference (SIU)*, 2015. IEEE, Malatya, Turkey.
    Communications,Signal Processing and Networking (WiSPNET), 2016. IEEE, Chennai, India.
46. N. Walia, H. Singh, S. K. Tiwari, and A. Sharma, "Design and identification of tuberculosis using fuzzy based decision support system," *Advances in Computer Science and Information Technology*, vol. 2, no. 3, pp. 57–62, 2015.
47. N. Walia, H. Singh, S. K. Tiwari, and A. Sharma, "A decision support system for tuberculosis diagnosability," *International Journal on Soft Computing*, vol. 6, no. 3, pp. 1–14, 2015.

48. C. F. CostaFilho, P.C. Levy, C.M. Xavier, M.G. Costa, L.B. Fujimoto, and J. Salem, "*Mycobacterium tuberculosis* recognition with conventional microscopy,"*2012 Annual International Conference of the IEEE Engineering in Medicine and Biology Society*, 2012. IEEE, San Diego, CA.
49. L. H. R. A. Évora, J. M. Seixas, and A. L. Kritski, "Artificial neural network models for diagnosis support of drug and multidrug resistant tuberculosis."*2015 Latin America Congress on Computational Intelligence (LA-CCI)*, 2015. IEEE, Curitiba, Brazil.
50. R.M. Nemeş, F. Mihăltan, R. Nedelcu, P. Postolache, M. Nitu, M.L. Baean, D. Todea, and D. C. Cojocaru, "Nicotinic withdrawal rate assessment by applying minimal advice and measurement of CO in the exhaled air in smoking patients with tuberculosis,"*2013 E-Health and Bioengineering Conference (EHB)*, 2013. IEEE, Iasi, Romania.
51. T. Uçar, A. Karahoca, and D. Karahoca, "Tuberculosis disease diagnosis by using adaptive neuro fuzzy inference system and rough sets," *Neural Computing and Applications*, vol. 23, no. 2, pp. 471–483, 2012.
52. E. M. Paul, B. Perumal, and M. Pallikonda Rajasekaran, "Filters used in X-ray chest images for initial stage tuberculosis detection,"*2018 International Conference on Inventive Research in Computing Applications (ICIRCA)*, 2018. IEEE, Coimbatore, India.
53. B. Castaneda, N. G. Aguilar, J. Ticona, D. Kanashiro, R. Lavarello, and L. Huaroto, "Automated tuberculosis screening using image processing tools," *2010 Pan American Health Care Exchanges*, 2010. IEEE, Lima, Peru.
54. B. Lenseigne, P. Brodin, H. Jeon, T. Christophe, and A. Genovesio, "Support vector machines for automatic detection of tuberculosis bacteria in confocal microscopy images," *2007 4th IEEE International Symposium on Biomedical Imaging: From Nano to Macro*, 2007. IEEE, Arlington, VA.

11 Applications of Artificial Intelligence in Detection and Treatment of COVID-19

Mangesh Pradeep Kulkarni, Rajesh Kumar, P. B. Vandana, Sagar, Tusara Kanta Behera, Sheetu Wadhwa, Gurvinder Singh, Pardeep Kumar Sharma, Deepika Sharma
Lovely Professional University

Cherry Bhargava
Symbiosis Institute of Technology

Sesha Sai Kiran Poluri
University of Greenwich

CONTENTS

11.1 INTRODUCTION

Inception of COVID-19 disease is a happening curse caused by the outbreak of Coronavirus covering almost the entire world. This pandemic condition has eaten up many lives, and still people are struggling between life and death every day. The early signs of disease were first noticed in Wuhan of China in December 2019, which, being highly contagious, spread across the globe. As a rescue strategy, many countries have made different policies against this disease for reducing the severity. Artificial intelligence (AI) is one such efficient policy, which can handle the situation better than any other aids. AI-powered drones have been used in surveillance and disinfectant activity. Moreover, the technology helps in predicting the disease state in early stages using imaging data like X-rays and CT (computed tomography) scan. These techniques are components of AI, which has been running successfully in various fields.

This chapter mainly focuses on the role of AI, which is used in the battle against COVID-19. The current advancement in AI is leading humans to upgrade their lives, which can be used in a significant way to reduce the pandemic impact. Through this manuscript, the readers will be able to understand different strategies based on AI that are applied in various healthcare sectors to fight effectively against COVID-19.

11.2 INCEPTION OF ARTIFICIAL INTELLIGENCE IN HEALTHCARE

AI in healthcare utilizes certain complex algorithms and software, which imitate the human intelligence in processing (Clancey & Shortliffe, 1984). During the 1960s and 1970s, researchers have worked on AI and produced an expert system, which was the first problem-solving program named Dendral. MYCIN was another system subsequent to Dendral. It was mainly applied in organic chemistry (Swartout, 1985). The extensive use of microcomputers with a unique network had started in the years of the 1980s and 1990s. In this era, researchers and developers recognized that AI in the healthcare promotes quality in visualizing data and helps in the expertise of physicians (Duda & Shortliffe, 1983).

11.2.1 Applications of AI in Healthcare

Radiology: Stanford created an algorithm that detects pneumonia at a specific site with a better average F1 metric than radiologists. Thus, the ability to interpret imaging results helps clinicians in the detection of minute changes in an image, which can be overlooked by the clinician accidentally (Rajpurkar et al., 2017).

Imaging: Recently, the advances in AI have eased the work to describe and evaluate the outcome of maxillofacial surgery or assess the cleft palate therapy regarding facial attractiveness (Patcas, Bernini et al., 2019; Patcas, Timofte et al., 2019).

In 2018, an article in *Annals of Oncology journal* had stated that AI could more accurately detect skin cancer, which refines the existing medication (Presse, 2018).

Disease Diagnosis: There are several AI techniques that are used for the diagnosis of a variety of diseases. Few are support vector machines, decision trees, neural networks etc. (Jiang et al., 2017). Demonstration of some specific functions in disease diagnosis is done by two different techniques, namely, artificial neural networks (ANN) (Bhargava, Banga, & Singh, 2016) and Bayesian networks (BN).

Thus, early classification and diagnosis of severe diseases like diabetes and cardiovascular diseases can be achieved by the development of machine learning models such as ANN and BN. Further, it is stated that ANN could more accurately classify these diseases as compared to BN (Alić, Gurbeta, & Badnjević, 2017).

Telehealth: Proliferation of Telemedicine has raised enormous applications of AI. Through AI, patients can be monitored easily with ease in communication between patients and physician that helps patients explain the symptoms better and makes the physician understand the case and diagnose it well. With AI, a patient can be monitored and assisted very well as compared to humans (Pacis, Subido Jr, & Bugtai, 2018).

Electronic Health Records (EHRs): These records are very helpful in digitalization and also in the spread of information in the healthcare industries. EHRs can be efficiently utilized only when AI tool is used, which can scan EHRs easily and predict the course of diseased person accurately (Häyrinen, Saranto, & Nykänen, 2008).

Drug Interactions: Natural language processing improvement such as algorithm development has enabled the identification of drug–drug interactions. Drug interactions, sometimes, are life-threatening to patients consuming multiple medications at a time (Cai et al., 2017). The role of AI in drug–drug interactions in tracking and generating an exact information of possible adverse effects has been highly appreciated (Christopoulou, Tran, Sahu, Miwa, & Ananiadou, 2020). Thus, it eases the work of doctors for the submission of reports on possible adverse reactions of medications to the organizations such as FDA Adverse Event Reporting System and WHO's VigiBase (Zhou, Miao, & He, 2018).

Creation of New Drugs: A drug molecule for the treatment of obsessive–compulsive disorder known as DSP-1181 was invented through AI and was accepted for the human trial. This was invented by the joint efforts of Exscientia and Sumitomo Dainippon Pharma, and the drug development has taken only 1 year, which, in general, takes 5 years.

Industry: Greater health data can be obtained when big companies merge with other companies, thus allowing an increment in the implementation of AI algorithms (Bhargava et al., 2020). Large companies are providing AI algorithms for aiming to process the data by finding a better clue (Panesar, 2019).

For example:

IBM: Watson Oncology is a technical approach at Memorial Sloan Kettering Cancer Center and Cleveland Clinic, which relies on AI applications for chronic disease treatments with CVS Health (USA).

Microsoft: By a partnership with Oregon Health and Science University (Knight Cancer Institute), it supports in investigation and medical research for the prediction of a most effective drug for cancer therapy.

Startups: Kheiron Medical has introduced a deep learning software, which can direct ways to study breast cancers in mammograms (Bluemke, 2018).

11.3 ARTIFICIAL INTELLIGENCE IN THE MANAGEMENT OF COVID-19

AI techniques have been involved in various areas related to Coronavirus pandemic, which include:

- AI in early detection and alert systems
- AI in tracking the patients along with predictions
- AI in diagnosis, treatment, and cure of the disease
- AI in obtaining the status and numbers related to disease using dashboards
- AI in social safety, surveillance, and prevention

11.3.1 AI in Early Detection and Alert Systems

BlueDot: A cluster of pneumonia patients was emerging around the fish market in Wuhan, China, as spotted by BlueDot (an AI system) on December 30, 2019 (Inn, 2020). After approximately 9 days, the condition was recognized, and the WHO declared warning statements (Inn, 2020). The attack was accredited as COVID-19 and later was declared as a pandemic outbreak by WHO, considering the spread and severity of the cases across the globe (Allam & Dhunny, 2019).

BlueDot is an organization that was launched in 2014 at Toronto, Canada, which involves a panel of highly qualified personnel like physicians, epidemiologists, veterinarians, software developers, and the data analysts along with scientists from different fields (Allam & Jones, 2020). The personnel utilized the natural language processing as a tool to generate artificial responses and further optimized to process Big Data in a limited time. This technology can grab information from various possible sources like digital media, global airline tickets, population demography, livestock health reports etc., and apply them while processing (Castro, McLaughlin, & Chivot, 2019).

Working of BlueDot: It has an extensive software, which gives service to locate, predict, and trace the spread of virus. BlueDot engine accumulates the data of over 150 diseases, and syndromes, which are registered throughout the world.

Besides providing official data, it can also extract the information of billions of the passengers traveling through various routes; human–animal population data; and information from the journalists, media, and healthcare workers.

It processes the information by classifying the data manually and creates a taxonomy for further learning activity. By giving a proper input, it provides a handy data on a specific topic or a case, and is also able to produce suitable traces for the needed investigations. It can also present recent or live updates to give an alert for the troublesome circumstances, thereby looking for a preventive action.

With respect to the COVID-19 attack, the BlueDot system has sensed and flagged Wuhan city rightly as the hub of the virus outbreak. It even anticipated the list of places like Bangkok, Tokyo, Phuket, Seoul, Taipei, and Singapore, as areas prone to develop infected conditions. This is not the first time; the involvement of this technology was evident; even in the previous years during Zika virus outbreak, this was involved therein.

Chatbot: In recent times, WhatsApp has emerged as a routine in everyone's life so WHO has selected WhatsApp and launched a chatbot, which gives a prior information regarding COVID-19. There are frequent updates in the tool that provides the latest news via audio or text method regarding the COVID-19. The users have options to share the views or opinions at any point of time regarding COVID-19 pandemic (Inn, 2020).

This approach developed easiness by maintaining transparency among billions of people all over the world to have a uniform relevant information on the disease.

This chatbot includes the most advanced information related to symptoms, preventive measures, and the difference between the symptoms of regular flu and COVID-19. Additionally, it provides the live updates on the count of Coronavirus sufferers to help the government, health workers, caretakers, and decision-makers, which helps in an efficient policymaking (Fadhil & Schiavo, 2019), (Klein, Kulp, & Sarcevic, 2018).

Aarogya Setu App: Ministry of Electronics and IT of Indian Government has developed an app, for making the citizens aware of the pandemic situation. The app includes the risk factors and preventive steps to avoid the infection. It helps users to undergo a self-assessment to know their own health status. This app is available on Google play and app store for android and iOS, respectively.

How the App Works: Aarogya Setu adopts Bluetooth-enabled tracking to keep the user informed with enough information, just in a case, if he/she comes closer to someone who is tested positive for COVID-19. The Bluetooth and live location features enable tracking of an individual location and generate a social graph, which shows the interaction with several people.

After the installation process is done, the user has to allow its Bluetooth and location sensors to activate and set the permissions on for continuous tracking. The app also conducts a survey regarding the COVID-19 and asks the various questions related to the personal symptoms. The report is forwarded to the government to keep them updated. While collecting data, the app also senses the infected person moving around and immediately warns the user to get isolated and safe. The data remains confidential with an access to the government and user himself.

11.4 ROLE OF AI IN TRACKING AND PREDICTION OF COVID-19

The pandemic circumstances of COVID-19 have reached to a critical level till date, which needs intellectual aids to answer the situation. After reaching the warning stages, different nations in collaboration with WHO decided to execute different strategies striving to rescue their position. To have these measures under control, the scientists utilized AI, which aided in the clear findings of the virus pathway.

Several countries across the globe have implemented these in response to the disastrous pandemic situation (Long & Ehrenfeld, 2020).

11.4.1 Machine Learning

The machine learning is one such strategy, which brings out the artificial intellect to solve the issue. This machine learning is something related to the data, which might be collected from various sources using different means. It is extended to a deep learning process where the system can think itself by grabbing the data from multiple networks. The primary objective of machine learning is to utilize the data in identifying the primary source of spread and for breaking the possible connections to terminate the chain. The process involves different algorithms, which respond based on the previous data saved in the library (Raj, Dewar, Palacios, Rabadan, & Wiggins, 2011).

11.4.2 BlueDot Technology

The Canadian government, as the earliest step, have introduced a Blue Dot technology, which is a machine learning approach, where the growing symptomatic cases of COVID-19 are compared with the existing data, to find the relative pattern. This technology employs the past data records and analyzes the similarity factors between them. This model demonstrates the extent of development, and the areas that are more affected. As an extension, comprehensive precognition can be carried out for visualizing the hotspots prone to Coronavirus, through which a considerable level of alertness can be created (Long & Ehrenfeld, 2020).

11.4.3 Spatial Analysis

By sourcing geographic information, the spread pattern of the virus can be easily hunted. This spatial analysis focuses on humans confined to a particular location. The inference of the Bayesian method facilitates the determination of infected ratio concerning time and space (Weblink1, 2020)

John Hopkins model is a platform, which presents live updates on the various epidemic scenarios in different countries (Weblink2, 2019).

National language processing is an AI tool, which uses different languages to encode the data and process them for output in the natural language. It converts the text into a structured format for further analyzing and displaying the results (Friedman, Rindflesch, & Corn, 2013).

Travel Data Collection is another vital tool to be considered as the cause of the spread that comes through migration. It is the ideal way to figure out the possible cases and help in early measurement of risks. This kind of analysis is represented in the form of a Sankey diagram designed by WHO, which plots the chain flow in multidimensional visions at various stages. For a better knowledge of the virus spread pattern, a graph analysis using historical data is employed, which recognizes the extent of the outbreak and notices where it is happening more. A network map has been developed by Singapore's coding academy to envision the outbreak influence in various parts of country. A screening system is implemented in the airports while exiting, where smart sensors are introduced. These sensors check for unusual body conditions to detect any positive signs of COVID-19 (Quilty, Clifford, Flasche, & Eggo, 2020).

11.4.4 Enter Telco Analytics

Involvement of electronic gadgets is a smart idea for individual tracking. The "Enter Telco Analytics" is a novel technology, which can collect the information of almost all the categories of people. Every electronic device is programmed to save few details of users, which will be tracked to monitor one's movement, action, and social behavior to check for abnormality. It utilizes different electronic gadgets like tablets, smartwatch, fit-band, smartphones, and other commonly accessible devices. Moreover, it is evident that on an average, each individual of this world carries at least one device with him/her. The smartphone is an elegant tool, as the software supports locating sensors, and fortunately, it is there in almost every pocket. It is not less than a library, which includes several pieces of information. So, tracking of one's smartphone may count the details from his name, location, contacts, and browsing history, and can even assess the behavior of a person. There is a process called sniffing, which can track the private information like browsing history, chats, and calls of an individual in emergency times. Usually, a regulatory body governs this, which is known as the General Data Protection Regulation, which is present in almost all the countries (Stopczynski et al., 2014; Valentino-DeVries, Singer, Keller, & Krolik, 2018).

11.4.5 Social Media

The advent of social media like Facebook, YouTube, WhatsApp, Instagram, and Twitter may serve in communicating the severity of the circumstances and bring consciousness among the viewers. Many programmers are developing trendy applications to detect the clues of infected people. Aarogya Setu is one of such trending apps in India, which does recognize the activity. Many more such applications are being developed, to strengthen the accessibility of needs and service to the public. The CCTV surveillance is probably a compelling idea to investigate the mode of transmission. Through it, the epidemic alerts are monitored, and an immediate preventive action can be scaled. The continuous enquiry of medical data may picturize the plot of epidemic projections (Grind, McMillan, & Wilde Mathews, 2020).

In contrast to the above-mentioned authentic applications, there are also a few false claims on AI being spread in the media. Through social media, people are posting their opinions as inference without placing proper evidence, which mislead the people and even create overexcitement and panic. Recently, a propaganda became viral on the internet that claimed the development of a biochip to track the Corona by Microsoft as reported by Reuters.

Despite many beneficial aspects, still a lot has to put forward in the utilization of AI to regulate the severity of the condition, instead of compromising the condition.

11.5 AI IN COVID-19 DIAGNOSIS

Detection of COVID-19 is the most crucial step to curb the transfer of the virus and avoid its progression to the successive stages of transmission. The major problem with this disease is its development period, which happens to be around 2–14 days. Being undetected for such a long period, the virus poses a greater level of risks, which is more likely to transmit the disease from one to too many. Thus, efficient diagnosis and isolation are the only keys to stay safe and prevent the spread of disease (Pokhrel, Hu, & Mao, 2020).

The testing can be primarily categorized into two types, i.e., detecting the presence of a virus or the viral proteins in the body, and looking for the presence of antibodies in the body produced in response to the disease.

11.5.1 Real-Time Reverse Transcriptase Polymerase Chain Reaction (rRT-PCR)

The real-time reverse transcriptase polymerase chain reaction (rRT-PCR) method is applied for the qualitative detection of a nucleic acid of SARS-CoV-2 from the specimens of the suspected population. This test can be done in two different formats, i.e., a single plex format in which three assays are done separately, and the other is the multiplex format involving a single reaction and a software setup. For the amplification of results, QuantStudio-7 software is used, and by observing the fluorescence activity (due to the dyes used in the test) at various cycles of the polymerase reaction, differentiation is noticed between the host and virus nucleic materials. In the same test, AI-mediated algorithms such as sparse re-scaled linear square regression (SRLSR), attribute reduction with multiobjective decomposition-ensemble optimizer (ARMED), gradient-boosted feature selection (GFS), and recursive feature elimination (RFE) together are employed for faster results and better confirmations. This test has primarily been used in the detection of infection in patients, but it is a time-consuming process.

So, to fulfill the needs of prompt diagnostic reports, several companies have started working on the novel test methods, a few of which have been briefly addressed here (Peng et al., 2020).

11.5.2 Antibody Detection Test

It is the test that uses antibodies to detect the presence of virus in the body. After the entry of disease in the body, the body develops antibodies against the viral antigens as a part of its immune response (Z. Li et al., 2020). The AI-based systems aid in establishing the identity of antibodies, which helps to trace the patterns of disease-causing virus and therefore its conformation (Guo et al., 2020).

11.5.3 Isothermal Nucleic Acid Amplification

US-FDA has recently approved an isothermal nucleic acid amplification testing, which is an AI-based approach. It is preferred instead of PCR for giving the positive outcomes within 5 minutes and the negative results in 18 minutes, and can be a good alternative to RT-PCR for increasing the testing efficiency (Weblink3).

11.5.4 CT Imaging Analysis

In this technique, the CT scan chest reports of pneumonia patients from August 2016 to February 2020 were collected and compared with COVID-19 patients by employing software U-net and COVnet, which scanned every detail and provided the inference (L. Li et al., 2020).

The radiological images obtained during the test were further processed by AI for quantification. The reported results show the accuracy of 98.2% and the specificity of 92.2% (Gozes et al., 2020).

Several researchers also recommended the use of AI in reading the CT scans of possible lung damage as it could reduce the time required with a great amount of accuracy. However, it was advised to get an RT-PCR test done later for the final confirmation of the disease.

11.5.5 Detection Using the Sensors of Smartphones

The present-day smartphones are embedded with several sensors such as the proximity, camera, microphone, temperature, fingerprint, inertial sensors, etc., for performing different operations. By employing these features for few diagnostic applications like usage of the microphone to record the pattern of coughing, similar parameters can be compiled and sent to a pre-developed, installed algorithm that can amplify and provide suggestions over the presence or absence of the infection.

This kind of analysis is preferred for the initial studies, to detect the traces of disease, and further processed for confirmation. In practical, however, there are higher chances of technical disturbances, and the reports may not be clear (Maghdid, Ghafoor, Sadiq, Curran, & Rabie, 2020).

11.6 AI IN THE TREATMENT OF COVID-19

With an increase in Corona victims day-by-day reaching great heights, a promising therapy is essential now, without costing more lives. In nearly 150 countries, several companies are struggling to bring the best. In this tough time, the utilization of AI seems like a door of hope. AI mediates the computer-aided drug design, which programs the structure of a drug and simulates thousands of molecules within a short span. For combating against COVID-19, a virus-suppressing tool is required.

Development and launch of new vaccine seem too tough a task to accomplish immediately; hence, attempts are being made to look back towards the existing vaccines or antiviral agents, i.e., their repurposing to get benefited out of it. For this, the historical background has been viewed for different diseases to recognize the pathogen having alike historical contagion. By going through the various findings, it has been observed that the present attacking strain is mildly getting neutralized by the

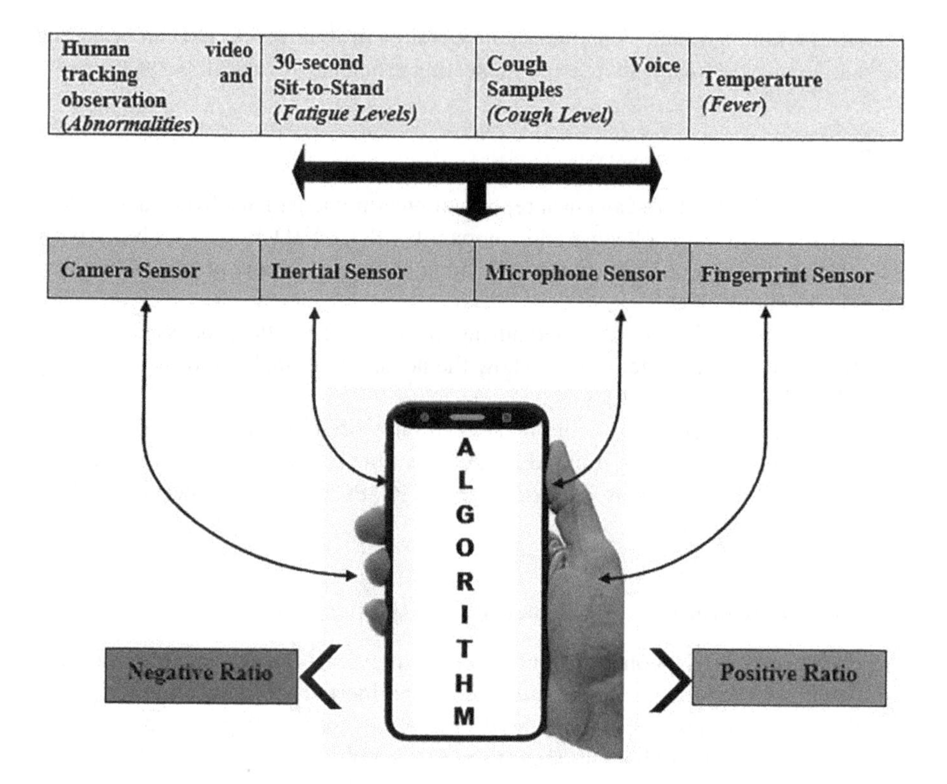

FIGURE 11.1 Detection using the sensors of smartphones.

malarial-suppressing drugs like hydroxychloroquine. Figure 11.1 shows the detection using the sensors of smartphones.

When this drug was targeted against Corona in the clinical trials, the responses were effective and showed a considerable impact in the recovering process. However, the recovery is quite lower in elderly patients in comparison with younger ones, which can also be considered insignificant (Colson, Rolain, Lagier, Brouqui, & Raoult, 2020).

The recent studies confirmed that the antiviral agent remdesivir manifested the promising outcomes when tried on the Corona victim. Remdesivir drug has been developed by Gilead Sciences using the computer-aided mechanism under AI. The US Government have conducted the clinical trials for this drug on the Corona-infected monkeys. While testing, the monkeys were categorized into two groups, in which one group were drugged and others are not (Wang et al., 2020).

When compared at post-stages, the breathing patterns of the two groups were dissimilar. The monkeys that consumed the drug had an efficient breathing pattern than the others. The experiment proved that the antiviral agent, remdesivir, had destroyed the virus to a certain extent. However, the trials are currently ongoing with this drug in the existing patients, to check the closest possibility and establish its potential as COVID-19 treatment strategy (Al-Tawfiq, Al-Homoud, & Memish, 2020).

The other encouraging medication which is under study to battle against the Coronavirus is gimsilumab, which is an artificially synthesized monoclonal antibody that may defeat the virus through immunotherapy. A pharmaceutical company Roivant Sciences conducted human trials for the first time, to determine the potency of gimsilumab against the Coronavirus. The drug has appeared effective during nonclinical studies, and currently, it has crossed through two stages of clinical studies. For satisfying the concerns of safety and tolerance, the drug has passed the phase 1 test. This drug aims to modulate the immune conditions, which cause a destructive mechanism on the virus (Weblink4, 2020).

11.7 AI IN MAINTENANCE OF THE AFFECTED AREAS AND DASHBOARD

When the rate of disease spread is rapid, the movement of information must be faster than the other cases. For providing the data, the dashboards are utilized, which provide the correct and relevant information on Coronavirus results from different parts of the world (Dong, Du, & Gardner, 2020). Some of the dashboards, which are presently in use, are given in the following subsections.

11.7.1 Johns Hopkins University Centre for Systems Science and Engineering Dashboard (JHU CSSE)

JHU CSSE dashboard has been developed by Lauren Gardner (an epidemiologist) and her team members, which is a leading dashboard by having nearly 140 million views and is also a platform for hundreds of articles and social media (Boulos & Geraghty, 2020).

JHU CSSE dashboard provides an interactive map that continuously locates the COVID-19-confirmed cases throughout the world along with the death and recovery cases. Viewers can have easy access to the updated data time to time. This software is controlled under five different authorities to show the collected data:

- World Health Organization
- US Centers for Disease Control and Prevention
- National Health Commission of the People's Republic of China
- European Centre for Disease Prevention and Control
- Chinese Online medical resources (Boulos & Geraghty, 2020)

The main limitation of this dashboard is that it does not store the visualized data of previous days. The panel only gives the timeline chart of a total number of the confirmed cases and recovered cases but the viewers are not able to retrieve and display the detailed map snapshots.

11.7.2 The World Health Organization (WHO) Dashboard

WHO, the main body that coordinates and directs the health globally, implemented measures to provide continuous surveillance for combating COVID-19. On January,

26, 2020, in China, it has prepared a dashboard ArcGIS for COVID-19, which marks, maps, and lists the total number of Corona-infected cases along with the deceased number of people.

Initially, a remarkable difference was observed between two dashboards of JSS CSSE and WHO, as both of them focused on different areas, where WHO relied on the laboratory-confirmed cases, while JHU CSSE claimed based on the diagnosis through symptoms and chest imaging. Later, both of these dashboards started working in sync and showing a similar number of the cases (Muhareb & Giacaman, 2020).

WHO panel displays a curve which depicts the infected ratio from date to date. The provided cumulative curve and the epidemic curve convey the vital information regarding the outbreak progression. The menu at the top-right corner presents additional important details about COVID-19 and an interactive map, which explains the COVID-19 context by WHO-monitored health emergencies (Boulos & Geraghty, 2020).

WHO dashboard is updated multiple times a day automatically by using ArcGIS GeoEvent server.

China Coronavirus Close Contact Detector App: As the government implemented the travel restriction and social gathering, so for initiating and monitoring the social distancing activity, the Government of China collaborated with National Health Commission and China Electronic Technology Group Corporation, and developed an app called "close contact detector." This app supervises the user if any infected person crosses from nearby and warns the user to get alert if the infecting person is close to him. This app utilizes the data from public authorities (data from flight, bus, and train booking) and can track the movement of an individual. Using its data, it checks whether the user has any close contact with any confirmed or suspected case. This app informs the user based on his/her location and recent movements.

This app can easily be accessed through the most popular mobile apps used in China, like Alipay, WeChat, and QQ (Boulos & Geraghty, 2020).

A voluntary system is also implemented in Guangzhou (China), which, in case a person is diagnosed positive to Coronavirus at post-stages of testing, enables to trace his/her transport routine to notify the related passengers who board the same public transport with the infected person. Since February 17, 2020, each station in Guangzhou displays a QR code, which passengers have to scan once when they board the train. Then, they have to fill a quick online form which appears on their mobile that includes info like their ID number, gender, starting station, and destination. This information is transferred to the government through an online mode, and if anyone confirms with positive symptoms of Coronavirus after testing, and has travel history in this app too, the fellow passengers are warned and traced easily (Boulos & Geraghty, 2020; Weblink5; Weblink6).

11.8 AI IN SOCIAL SAFETY/SURVEILLANCE/ PREVENTION OF COVID-19

As COVID-19 has grown to a pandemic level, the need for drones and robots has tremendously been increased owing to the need of implementing strict social distancing measures. As in cases of contagious disease, it is a challenging task to restrict the people from a social gathering; hence, monitoring plays a crucial role and drones are perfect for such a role. Drones continuously monitor the individuals for wearing of masks or avoiding public gathering and even disinfecting activities taking place in public places.

A Shenzhen-based company called Micro Multi Copter used their drones to carry medicines and quarantine materials throughout the city and also for patient care services without involving healthcare providers. This helped healthcare workers to reduce the risk of spread. Moreover, they can also be used in cleaning and sterilization activities in infected wards. In catering industry center, Pudu Technology employed their robots in 40 hospitals for cleaning and sterilizing purposes.

COVID-19 is a SARS, which is highly contagious virus, that needs a continuous surveillance for controlling the spread. The main reason for COVID-19 becoming a global threat is human migration. Canadian AI startup BlueDot (a Toronto-based startup) extensively used machine learning and natural language processing to recognize the virus, to track it, and to report the spread of COVID-19 much quicker than WHO and US Center for Disease Control and Prevention. By using this technology, zoonoses can be foretold in the future, considering climatic shifts and human activity as variables.

Stallion. AI, which is a Canadian-based AI Research and Development Company, attached its natural language processing capabilities to a "multi-lingual virtual healthcare agent," which can answer the questions related to COVID-19, and provide reliable information and clear guidelines too. It can also recommend the protection measures, look for symptoms, and monitor them. It advises the individuals whether they have to self-isolate at home or to seek hospital screening (Weblink7).

Another AI-based software known as InferVISION informs that the patients having pneumonia symptoms are susceptible to COVID-19. The AI system is powered by NVIDIA's Clara SDKs, which is employed in imaging and genomics application and optimized for healthcare (McCall, 2020).

Researchers at National Tsing Hua University and researchers at Harvard University's School of Public Health in Taiwan are working with Facebook to track the travel history of infected people (Weblink8).

For taking preventive measures against COVID-19, some companies are offering free online sessions and webinars, while some are marketing the essential tools (Weblink9).

In Israel, two hospitals managed the COVID-19 infection by reducing the exposure of healthcare workers to infected patients. CLEW (AI-powered predictive analytics company) has taken out a solution, i.e., CLEW ICU (Tele-ICU), which was exercised at Sheba Medical Center and Ichilov Hospital at Tel Aviv Sourasky Medical Center.

It used AI to analyze the required expansion of the intensive care unit (ICU) capacity and also the resources exponentially. The algorithms had been trained for the advanced identification of respiratory deterioration, predicting the severity of disease using a remote command center. These large ICU units may use telemedicine technology for assisting the patients from remote areas by a centralized command. Thus, CLEW affirms that its machine learning models ease the work in ICU, overall workload, and also reduce the exposure of healthcare providers to the infected patients (Weblink10).

Smart Disinfection and Sanitation Tunnel: They are designed to provide disinfection and a maximum protection to the people by passing through a tunnel (Figure 11.2) for 15 seconds.

These can help the community to fight against COVID-19. The demonstration of how one undergoes sanitization in a tunnel is shown in Figure 11.3.

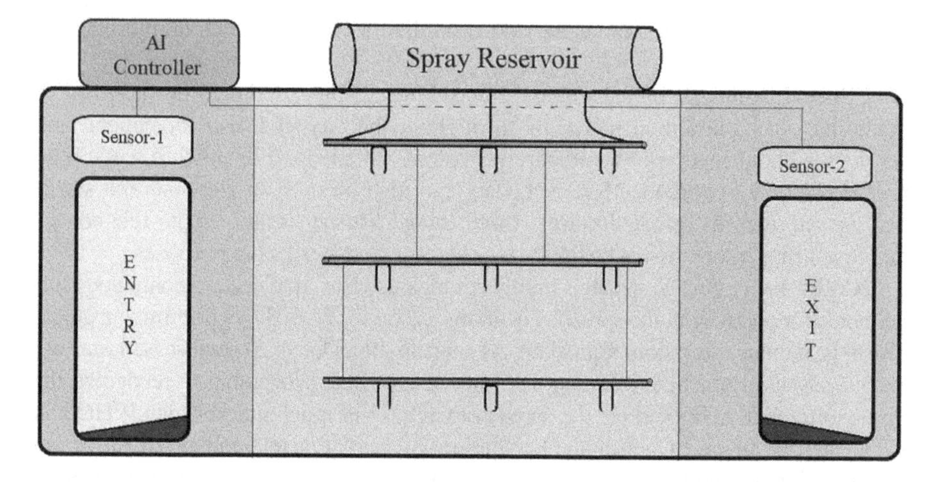

FIGURE 11.2 Schematic diagram of sanitization tunnel.

FIGURE 11.3 Demonstration of how one undergoes sanitization in a tunnel.

Its prototype was installed in Tamil Nadu State of India at a market where 1 ppm solution of sodium hypochlorite was sprayed on the people passing through the tunnel to sanitize them (Weblink11).

11.9 CONCLUSION

AI is becoming a vital tool in the healthcare sector and has still a lot to offer in the future. The earlier efforts in the diagnosis and treatment of diseases didn't prove much convincing; however, gradually AI has started making its impact in these domains as well. Though its role in current pandemic COVID-19 is also in its infancy, still it seems to hold a promise to be a part of an era wherein all the sectors would essentially involve its applications. A serious concern associated with AI is the privacy of data. Healthcare framework would require to continue the AI-based surveillance of citizens even after pandemic is over, and thus, concern about data privacy seems justified. So, it becomes crucial for the regulatory authorities to take an utmost care in data handling and stop its misuse.

REFERENCES

Al-Tawfiq, J. A., Al-Homoud, A. H., & Memish, Z. A. (2020). Remdesivir as a possible therapeutic option for the COVID-19. *Travel Medicine and Infectious Disease*. doi: 10.1016/j.tmaid.2020.101615.

Alić, B., Gurbeta, L., & Badnjević, A. (2017). Machine learning techniques for classification of diabetes and cardiovascular diseases. Paper presented at the *2017 6th Mediterranean Conference on Embedded Computing (MECO)*, Bar, Montenegro.

Allam, Z., & Dhunny, Z. A. (2019). On big data, artificial intelligence and smart cities. *Cities, 89*, 80–91.

Allam, Z., & Jones, D. S. (2020). On the coronavirus (COVID-19) outbreak and the smart city network: universal data sharing standards coupled with artificial intelligence (AI) to benefit urban health monitoring and management. In Healthcare (Vol. 8, No. 1, p. 46). Multidisciplinary Digital Publishing Institute.

Bhargava, C., Banga, V. K., & Singh, Y. (2016). Improved reliability, accuracy and failure minimization of spacecraft satellite architecture by using fuzzy logic. *International Journal of Control Theory and Applications, 9*(41), 523–532.

Bhargava, C., Sharma, P. K., Senthilkumar, M., Padmanaban, S., Ramachandaramurthy, V. K., Leonowicz, Z., . . . Mitolo, M. (2020). Review of health prognostics and condition monitoring of electronic components. *IEEE Access, 8*, 75163–75183.

Bluemke, D. A. (2018). Radiology in 2018: are you working with AI or being replaced by AI? *Radiology, 287*(2), 365–366.

Boulos, M. N. K., & Geraghty, E. M. (2020). Geographical tracking and mapping of coronavirus disease COVID-19/severe acute respiratory syndrome coronavirus 2 (SARS-CoV-2) epidemic and associated events around the world: how 21st century GIS technologies are supporting the global fight against outbreaks and epidemics. *International Journal of Health Geographics, 19*, 8.

Cai, R., Liu, M., Hu, Y., Melton, B. L., Matheny, M. E., Xu, H., . . . Waitman, L. R. (2017). Identification of adverse drug-drug interactions through causal association rule discovery from spontaneous adverse event reports. *Artificial Intelligence in Medicine, 76*, 7–15.

Castro, D., McLaughlin, M., & Chivot, E. (2019). *Who is winning the AI race: China, the EU, or the United States.* Center for Data Innovation, August, 19.

Christopoulou, F., Tran, T. T., Sahu, S. K., Miwa, M., & Ananiadou, S. (2020). Adverse drug events and medication relation extraction in electronic health records with ensemble deep learning methods. *Journal of the American Medical Informatics Association, 27*(1), 39–46.

Clancey, W. J., & Shortliffe, E. H. (1984). *Readings in Medical Artificial Intelligence: The First Decade*. Boston, MA: Addison-Wesley Longman Publishing Co., Inc.

Colson, P., Rolain, J.-M., Lagier, J.-C., Brouqui, P., & Raoult, D. (2020). Chloroquine and hydroxychloroquine as available weapons to fight COVID-19. *International Journal of Antimicrobial Agents, 55*, 105932. doi: 10.1016/j.ijantimicag.2020.105932.

Dong, E., Du, H., & Gardner, L. (2020). An interactive web-based dashboard to track COVID-19 in real time. *The Lancet Infectious Diseases, 20*, 533–534.

Duda, R. O., & Shortliffe, E. H. (1983). Expert systems research. *Science, 220*(4594), 261–268.

Fadhil, A., & Schiavo, G. (2019). Designing for health chatbots. arXiv preprint arXiv:1902.09022.

Friedman, C., Rindflesch, T. C., & Corn, M. (2013). Natural language processing: state of the art and prospects for significant progress, a workshop sponsored by the National Library of Medicine. *Journal of Biomedical Informatics, 46*(5), 765–773.

Gozes, O., Frid-Adar, M., Greenspan, H., Browning, P. D., Zhang, H., Ji, W., . . . Siegel, E. (2020). Rapid AI development cycle for the coronavirus (covid-19) pandemic: initial results for automated detection & patient monitoring using deep learning CT image analysis. arXiv preprint arXiv:2003.05037.

Grind, K., McMillan, R., & Wilde Mathews, A. (2020). To track virus, governments weigh surveillance tools that push privacy limits. *Wall Street Journal*. https://www.wsj.com/articles/to-track-virus-governments-weigh-surveillance-tools-that-pushprivacy-limits-11584479841.

Guo, L., Ren, L., Yang, S., Xiao, M., Chang, D., Yang, F., . . . Xiao, Y. (2020). Profiling early humoral response to diagnose novel coronavirus disease (COVID-19). *Clinical Infectious Diseases, 71*, 778–785.

Häyrinen, K., Saranto, K., & Nykänen, P. (2008). Definition, structure, content, use and impacts of electronic health records: a review of the research literature. *International journal of Medical Informatics, 77*(5), 291–304.

Inn, T. L. (2020). *Smart City Technologies Take on COVID-19*. Malaysia: World Health. https://penanginstitute.org/wp-content/uploads/2020/03/27_03_2020_TLI_download.pdf

Jiang, F., Jiang, Y., Zhi, H., Dong, Y., Li, H., Ma, S., . . . Wang, Y. (2017). Artificial intelligence in healthcare: past, present and future. *Stroke and Vascular Neurology, 2*(4), 230–243.

Klein, A., Kulp, L., & Sarcevic, A. (2018). Designing and optimizing digital applications for medical emergencies. *Extended Abstracts on Human Factors in Computing Systems*. CHI Conference, 2018, LBW588. doi: 10.1145/3170427.3188678.

Li, L., Qin, L., Xu, Z., Yin, Y., Wang, X., Kong, B., . . . Song, Q. (2020). Artificial intelligence distinguishes covid-19 from community acquired pneumonia on chest ct. *Radiology, 3*, 200905.

Li, Z., Yi, Y., Luo, X., Xiong, N., Liu, Y., Li, S., . . . Chen, W. (2020). Development and clinical application of a rapid IgM-IgG combined antibody test for SARS-CoV-2 infection diagnosis. *Journal of Medical Virology, 92*, 1518–1524.

Long, J. B., & Ehrenfeld, J. M. (2020). The role of augmented intelligence (AI) in detecting and preventing the spread of novel coronavirus. *Journal of Medical Systems, 44*, 59.

Maghdid, H. S., Ghafoor, K. Z., Sadiq, A. S., Curran, K., & Rabie, K. (2020). A novel AI-enabled framework to diagnose coronavirus COVID 19 using smartphone embedded sensors: design study. arXiv preprint arXiv:2003.07434.

McCall, B. (2020). COVID-19 and artificial intelligence: protecting health-care workers and curbing the spread. *The Lancet Digital Health, 2*(4), e166–e167.

Muhareb, R., & Giacaman, R. (2020). Tracking COVID-19 responsibly. *The Lancet*. doi: 10.1016/S0140-6736(20)30693-0.

Pacis, D. M. M., Subido Jr, E. D., & Bugtai, N. T. (2018). Trends in telemedicine utilizing artificial intelligence. In *AIP conference proceedings* (Vol. 1933, No. 1, p. 040009). AIP Publishing LLC..

Panesar, A. (2019). *Machine Learning and AI for Healthcare*. Berkeley, CA: Springer.

Patcas, R., Bernini, D. A., Volokitin, A., Agustsson, E., Rothe, R., & Timofte, R. (2019). Applying artificial intelligence to assess the impact of orthognathic treatment on facial attractiveness and estimated age. *International Journal of Oral and Maxillofacial Surgery, 48*(1), 77–83.

Patcas, R., Timofte, R., Volokitin, A., Agustsson, E., Eliades, T., Eichenberger, M., & Bornstein, M. M. (2019). Facial attractiveness of cleft patients: a direct comparison between artificial-intelligence-based scoring and conventional rater groups. *European Journal of Orthodontics, 41*(4), 428–433.

Peng, M., Yang, J., Shi, Q., Ying, L., Zhu, H., Zhu, G., . . . Wang, J. (2020). Artificial intelligence application in COVID-19 diagnosis and prediction. https://ssrn.com/abstract=3541119 or http://dx.doi.org/10.2139/ssrn.3541119

Pokhrel, P., Hu, C., & Mao, H. (2020). Detecting the coronavirus (COVID-19). *ACS Sensors, 5*(8), 2283–2296.

Presse, A. F. (2018). Computer learns to detect skin cancer more accurately than doctors. *The Guardian*, 29.

Quilty, B. J., Clifford, S., Flasche, S., & Eggo, R. M. (2020). Effectiveness of airport screening at detecting travellers infected with novel coronavirus (2019-nCoV). *Eurosurveillance, 25*(5), 2000080.

Raj, A., Dewar, M., Palacios, G., Rabadan, R., & Wiggins, C. H. (2011). Identifying hosts of families of viruses: a machine learning approach. *PLoS One, 6*(12), e27631.

Rajpurkar, P., Irvin, J., Zhu, K., Yang, B., Mehta, H., Duan, T., . . . Shpanskaya, K. (2017). CheXNet: radiologist-level pneumonia detection on chest X-rays with deep learning. arXiv preprint arXiv:1711.05225.

Stopczynski, A., Sekara, V., Sapiezynski, P., Cuttone, A., Madsen, M. M., Larsen, J. E., & Lehmann, S. (2014). Measuring large-scale social networks with high resolution. *PLoS One, 9*(4), e95978.

Swartout, W. R. (1985). *Rule-Based Expert Systems: The Mycin Experiments of the Stanford Heuristic Programming Project*: BG Buchanan and EH Shortliffe (Addison-Wesley, Reading, MA, 1984); 702 p, $40.50.

Valentino-DeVries, J., Singer, N., Keller, M. H., & Krolik, A. (2018). Your apps know where you were last night, and they're not keeping it secret. *The New York Times*, 10.

Wang, M., Cao, R., Zhang, L., Yang, X., Liu, J., Xu, M., . . . Xiao, G. (2020). Remdesivir and chloroquine effectively inhibit the recently emerged novel coronavirus (2019-nCoV) in vitro. *Cell Research, 30*(3), 269–271.

Weblink1 (March 13, 2020). Geographic information systems/science: spatial analysis & modelling. Retrieved from https://researchguides.dartmouth.edu/gis/spatialanalysis.

Weblink2 (January 28, 2019). Retrieved from https://coronavirus.jhu.edu/.

Weblink3 (March 27, 2020). Update on COVID-19 in vitro diagnostics listed by National Regulatory Authorities in IMDRF jurisdictions. Retrieved from https://www.who.int/diagnostics_laboratory/200327_imdrf_covid19_listing_update_27_march_2020.pdf.

Weblink4. (April 16, 2020). Roivant starts gimsilumab dosing in COVID-19 trial. Retrieved from https://www.clinicaltrialsarena.com/news/roivant-gimsilumab-covid-19-trial/.

Weblink5 (April 16, 2020). WHO coronavirus disease (COVID-19) pandemic. Retrieved from https://www.who.int/emergencies/diseases/novel-coronavirus-2019.

Weblink6 (March 11, 2020). WHO COVID-19 dashboards. Retrieved from http://healthcybermap.org/WHO_COVID19/#1.

Weblink7 (March 30, 2020). How artificial intelligence is helping fight the COVID-19 pandemic. Retrieved from https://www.entrepreneur.com/article/348368.

Weblink8 (March 24, 2020). AI versus COVID-19: a soldier we did not know we need. Retrieved from https://www.analyticsinsight.net/ai-versus-covid-19-a-soldier-we-did-not-know-we-need/.

Weblink9 (March 31, 2020). Hospitals deploy AI tools to detect COVID-19 on chest scans. Retrieved from https://spectrum.ieee.org/the-human-os/biomedical/imaging/hospitals-deploy-ai-tools-detect-covid19-chest-scans.

Weblink10 (March 27, 2020). Artificial intelligence holds promise in improving revenue cycle management in healthcare. Retrieved from https://www.healthcareitnews.com/news/europe/two-israeli-hospitals-launch-ai-based-tele-icu-support-covid-19-patients.

Weblink11 (April 05, 2020). India's disinfectant tunnels – emerging strategies to combat coronavirus. Retrieved from https://www.investindia.gov.in/team-india-blogs/indias-disinfectant-tunnels-emerging-strategies-combat-coronavirus.

Zhou, D., Miao, L., & He, Y. (2018). Position-aware deep multi-task learning for drug–drug interaction extraction. *Artificial Intelligence in Medicine, 87*, 1–8.

12 Internet of Things *Powered Artificial Intelligence Using Microsoft Azure Platform*

Ranbir Singh and Amiya Kumar Dash
BML Munjal University

Ravinder Kumar
Lovely Professional University

Anand Bewoor
Cummins College of Engineering for Women

Ashwini Kumar
National Institute of Technology

CONTENTS

12.1 INTRODUCTION

According to Bernard Marr [1], the two most powerful technologies today are Internet of Things and artificial intelligence (AI); a combination of them is AIoT (the artificial intelligence of Internet of Things) [1]. This is one of the deadliest combinations forever. Both Internet of Things and AI seem to be made for each other and incomplete without each other. New opportunities and markets are appearing with a combination of AI and IoT [2], which will change our future in health care, marketing, manufacturing, banking, and each and every field of our life. It is forecasted by Business Insider that by 2026, installation of more than 64 billion IoT devices is expected worldwide, which was about 10 billion in 2018; about $ 900 billion investment is expected for smart city solutions annually by 2023. Revenue of Intel's IoT Group was 920 million USD in the fourth quarter of 2019 [3]. Technically, it is expected that low-power, wide-area networks (LPWANs) will give the global coverage by 2022. LPWAN is the cheapest and least power-consuming device for the long-range communications among IoT devices in large number [4]. IoT-enabled AI will be the future of the technology [5]. Market of AIoT devices will approach to $ 26.2 billion by 2023 [6]. This chapter is drafted around working with AI-powered IoT.

Learning system capable to emulate manual tasks without any human intervention is termed as an intelligent system. Hardware is termed as machine (with microprocessor). Learning capabilities of the system is termed as machine learning (ML)/ deep learning. For ML, it should be feed with structured/organized data (via IoT), from which information is derived for decision-making called as AI. IoT is a network of sensors, microcontroller, and cloud essentially connected via internet to collect and exchange data.

Internet of Things: A network of embedded physical devices to sense, interact, and communicate with other devices in the network.

Artificial Intelligence: Giving an ability of human intelligence to the machines.

Machine Learning: Data analysis approach to automate analytical modeling with iterative learning algorithms to find hidden insights without explicit programming

Deep Learning: ML with Big Data with a large number of independent process variables

Data is typically generated from a small amount of few KB to several hundred GB per unit time. If data is generated in small amounts, a stand-alone computer/memory device is sufficient to store the data for few hours to few days. If data is generated in such large amounts that a stand-alone computer/memory device cannot handle even per unit time data, it is called as Big Data. Data of the order of several GB to TB generated per unit time (say 5 GB/min, 100 GB/hour etc.) is referred as Big Data. There are three parameters of Big Data, namely, volume, velocity, and variety. For data to be big, a huge volume of data (from several GB to several TB) should be produced with a high velocity (of several hundreds to thousands of GB data within sec/min/hours) with a variation in its type (variable and unstructured mixup of the data: text, audio, pictures, video).

12.2 COMPUTING REQUIREMENTS

Vertical computing using Ram and memory of one CPU becomes incapable to deal with the produced data. Thus, for BIG DATA ANALYSIS, we need PARALLEL COMPUTING instead of vertical computing (one computer). PARALLEL COMPUTING is also called as HORIZONTAL COMPUTING/CLUSTER COMPUTING. A number of CPUs can be networked, or virtual machines (cloud) can be architectured for horizontal computing.

In the network/architecture, one machine acts as a master and the other as slave machine. Master machine distributes data to the slave, which on completion of analysis submits data back to the master machine which then produces/submits/generates the final results. In addition to several software packages for networking and control, transfer and management of data and results, we need to have several thousands of storage space for data storage and processing. There are two possible ways of doing it. Either we can network several number of CPUs together (initial investment is required to have such numbers of CPUs) in the required topology to generate several thousands of TB memory or we need to use cloud services for the purpose. Our software requirement reduces if we select to use cloud memory, as many required software packages are part of cloud services. Azure, IBM, and AWS are common among CLOUD SERVICES.

12.3 REAL-TIME DATA ANALYSIS

Jupiter/Pandas/Anaconda/python IDLE cannot perform a real-time data analysis and parallel computing. So, we need Big Data technology/concept to use parallel computing. The major Big Data tools are Hadoop and Apache spark.

Computing is done on RAM in Apache Spark and on disk in HADOOP. For the same amount of data, the speed of data handling/processing in APACHE SPARK is about 100 times faster/higher than that in HADOOP. Apache Spark can perform a real-time parallel computing/processing but Hadoop can't. Programming in APACHE SPARK is easily compared (as it supports different APIs) to HADOOP's map reducing. There is no need to install any third-party software for parallel computing in APACHE SPARK, which is otherwise required in HADOOP.

Focusing on Azure cloud platform, AZURE DATA BRICKS and AZURE HD INSIGHT are the two common parallel computing Azure cloud services/platforms for Big Data analysis. The difference among the two is that AZURE DATA BRICKS uses Apache Spark with spark core, whereas AZURE HD INSIGHT uses Apache Spark with Hadoop core. AZURE HD INSIGHT supports both Hadoop and Apache Spark, whereas AZURE DATA BRICKS supports Apache Spark only. Also, one can use API of Scala, python, and Java with Apache Spark.

In PARALLEL COMPUTING, software packages are required for the following:

- **Data Distribution**: Hadoop distributed file system (HDFS) is mostly used
- **Data Injection**
- **Cluster Management**: Pi Spark, Map reduce

Block chain is finding its increased application in Big Data analysis to ensure data security in networking and data sharing.

There are five important phases of Big Data analysis:

- Application
- Architecture
- Basic information
- Implementation
- Advancements

12.4 AIOT: INTEGRATION OF IOT & AI ON MICROSOFT AZURE PLATFORM

For Big Data analysis with IoT on a cloud platform, we need:

- Sensors
- Microcontrollers like Rpi computer/Node MCU/Arduino UNO/Ram controller etc.
- Parallel computing service [Cloud services] and
- Secure communications among the three.

To work with a cloud platform, we need to develop an IoT hub and a parallel computing platform on the cloud service. IoT hub is the point on the cloud platform with which the microcontroller to which sensors are feeding continuous data communicates. It is the communicator between the parallel computing platform and the microcontroller. It also provides security to the data. We need to prepare three modules in IoT hub: first, we need to generate an IoT device to get name of the device on the IoT hub on Azure Portal; second, we share access policies to generate SAS token for SSL security and lastly develop build-in endpoints to generate event hub-compatible endpoints.

IoT hub communicates the microcontroller data to the parallel computing platform (Azure Bricks) with security, i.e., SSL security with encrypted passwords, where data analysis is done on Big Data. Notebook (Apache Spark core) is used in azure bricks cluster for data analysis/ML/deep learning. Using techniques like neural networks with the data, AI systems are then developed.

Figure 12.1 represents the symbolic diagram of data analysis using IoT devices and Microsoft Azure platform.

A sensor module collects the data. This module is having hardware security. Its inputs and outputs cannot be tempered. This untampered data is fed to a microcontroller—here RPi computer. In a microcontroller, we need to write a program to fetch the sensor data into microcontroller and send it to a cloud platform. To send it to a cloud platform, microcontroller should always be connected with internet. Hence, microcontroller needs firmware security. Its hardware, software, & i/o's should not be tampered without permission and are secured by password-encrypted protections.

Machine-to-machine typically uses mqtt protocol (uses port no 1883 without SSL & 8883 with SSL) for data communication. It is because https (uses 443 or 80 no. port) is based on client and server-based architecture so it cannot transfer the real-time data: Advanced Message Queuing Protocol (AQMP) is slow compared to mqtt because of its higher security. Similarly, other communication protocols such as Constrained Application Protocol (CoAP), Extensible Messaging & Presence Protocol (XMPP), Data Distribution Service (DDS), and lightweight M2M (LwM2M) have their own limitations.

Usually, there are four types of security in IoT with AI and ML:

- Hardware security
- Network security
- Data security
- Cloud security

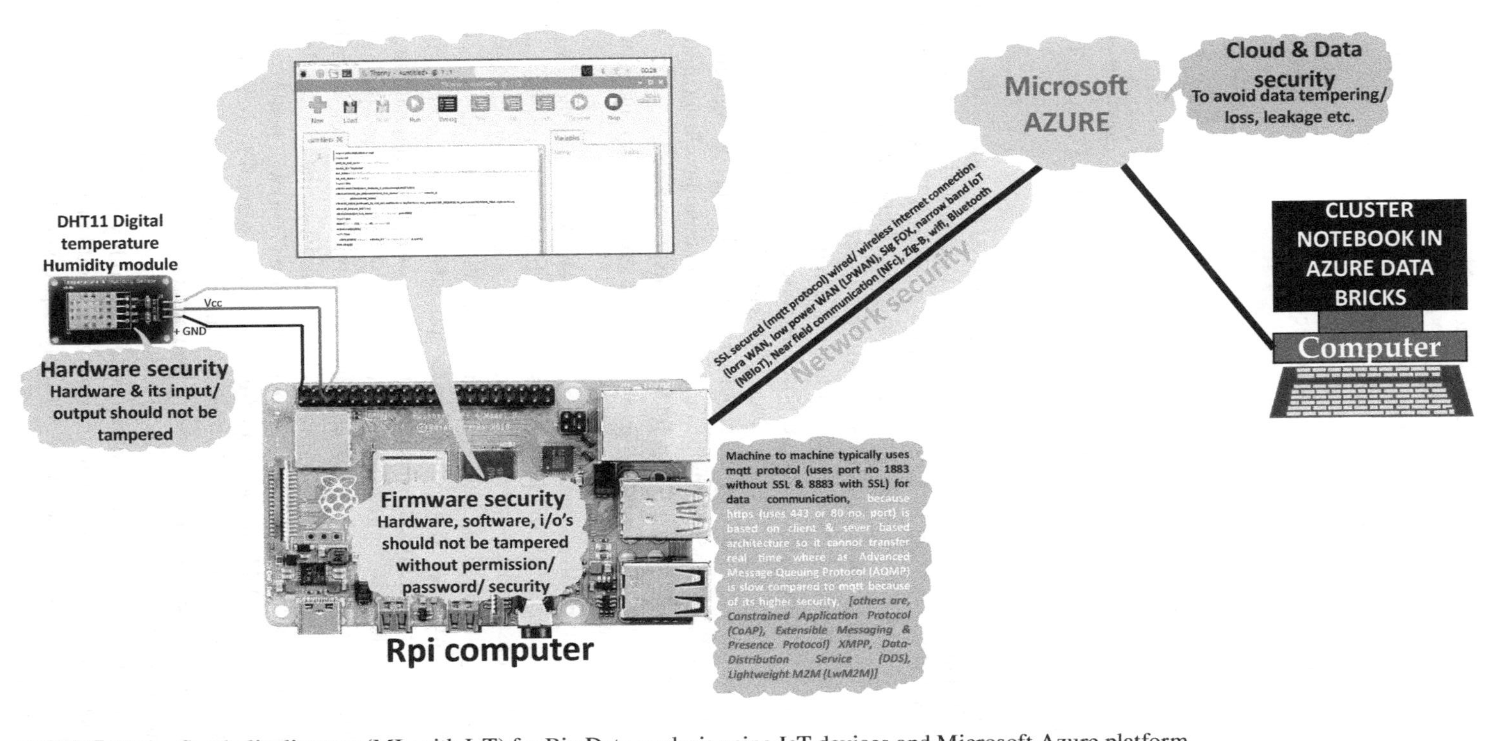

FIGURE 12.1 Symbolic diagram (ML with IoT) for Big Data analysis using IoT devices and Microsoft Azure platform.

Always use hardware(s) with security certifications. Network security is provided by SSL security by the encrypted connection. X509 in awsIoT and SAS token in azure IoT are the security certificates issued by SSL for network security. Port no for SSL security is 8883. Device explorer twin is the Microsoft tool with which we can manage IoT devices on azure. We need to attach CA root certificate for the purpose. SSL security is enabled as client Transport Layer Security (TLS) by SAS token by creating CA root certificate as follows:

- *client.tls_set*
- *ca_certs=path_to_root_cert,*
- *certfile=None,*
- *keyfile=None,*
- *cert_reqs=ssl.CERT_REQUIRED*,tls_version=ssl.PROTOCOL_TLSv1, ciphers=None

Now this data needs to be transferred from Rpi (microcontroller) to the cloud platform. For the purpose, we need network security. SSL secured (mqtt protocol) wired/wireless internet connection, lora WAN, low-power WAN (LPWAN), Sig FOX, narrow band IoT (NBIoT), near-field communication (NFc), Zig-B, Wi-Fi, and Bluetooth can be used for secure data transmission. All these mediums provide the required level of security for data transfer over the internet (wired/wireless). Industry typically uses Industrial Ethernet (IE) for the purpose (because of TCP/IP security).

Now when data is transferred to cloud platform, cloud and data security came into picture, which is provided by the cloud service provider (as per agreements) like Azure, IBM, AWS, or others. Finally, data analysis is done on cluster notebook in Azure Data Bricks with which AI applications are then developed.

Figure 12.2 presents the block diagram for mqtt protocol working showing the publisher, broker, and subscriber concepts.

Mqtt protocol typically works on the publisher–broker–subscriber fundamentals. One who generates data/request is termed as publisher. Mqtt broker stores the received data. Subscriber is one who seeks/needs this data. If the sensor is generating data and feeding to cloud (IoT hub) via a microcontroller, then the sensor is the publisher. If you are working on the computer for some data analysis work, you need to fetch data from the mqtt broker, which means you ae subscriber and mqtt broker will now be the publisher (as it already has that data in its memory) and publishes data for you. After analysis, if you need to control some actuator at the sensor end via the same or other microcontroller, then you will be the publisher, the cloud is the broker, and the microcontroller (fetch data via programs written in it) will be the subscriber. Some brokers are available as open source/free (within some limits), and most are paid. broker.hive.mq.com is one of the free available brokers. Always remember that data security might be a concern in case of free broker.

First, we need to write program in Rpi computer (microcontroller) to collect sensor data via DHT11 (generates temperature and humidity measurements data) and communicate it to Azure Bricks Cluster on Azure Cloud platform via some broker. Figure 12.3 presents the program architecture for Rpi computer to communicate sensor data to broker.

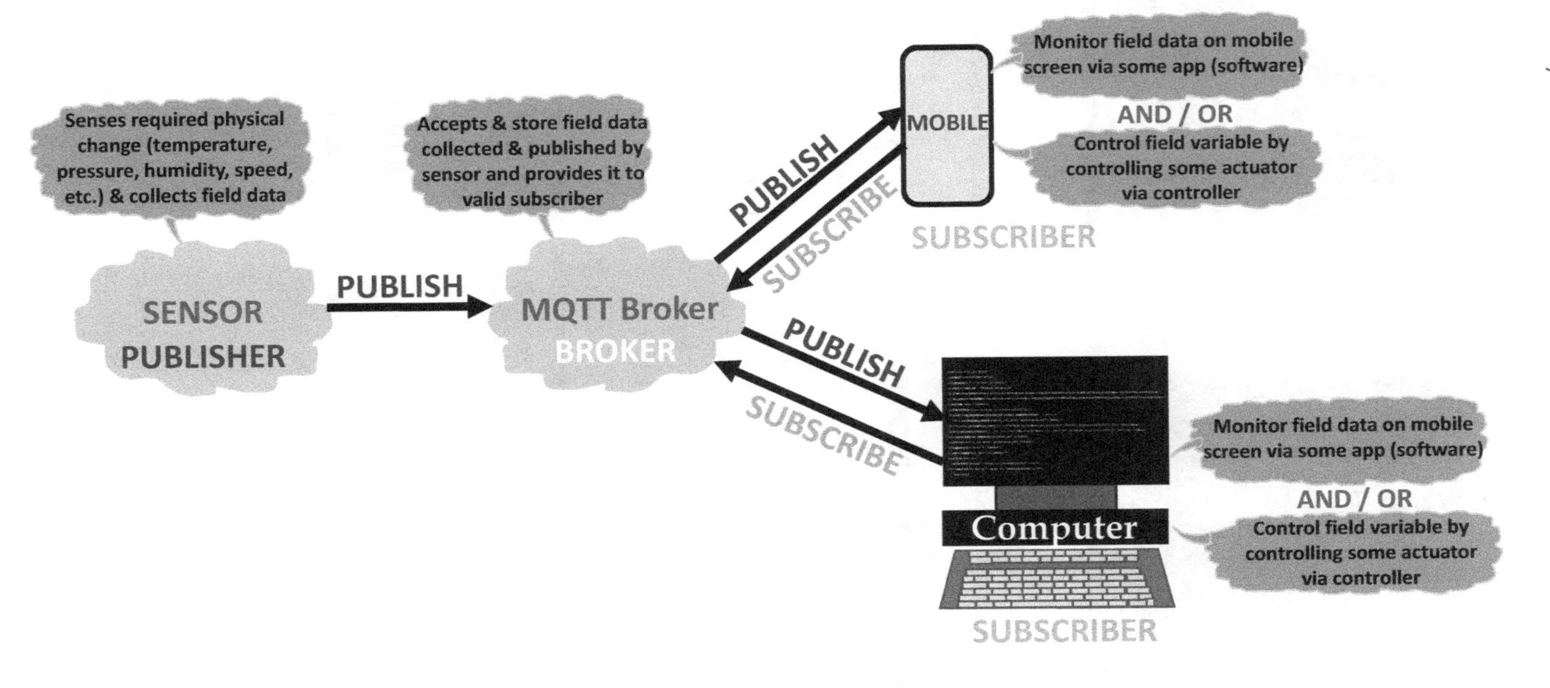

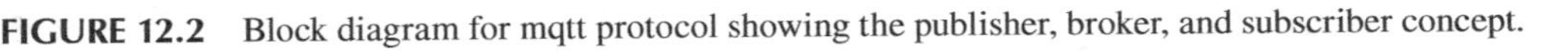

FIGURE 12.2 Block diagram for mqtt protocol showing the publisher, broker, and subscriber concept.

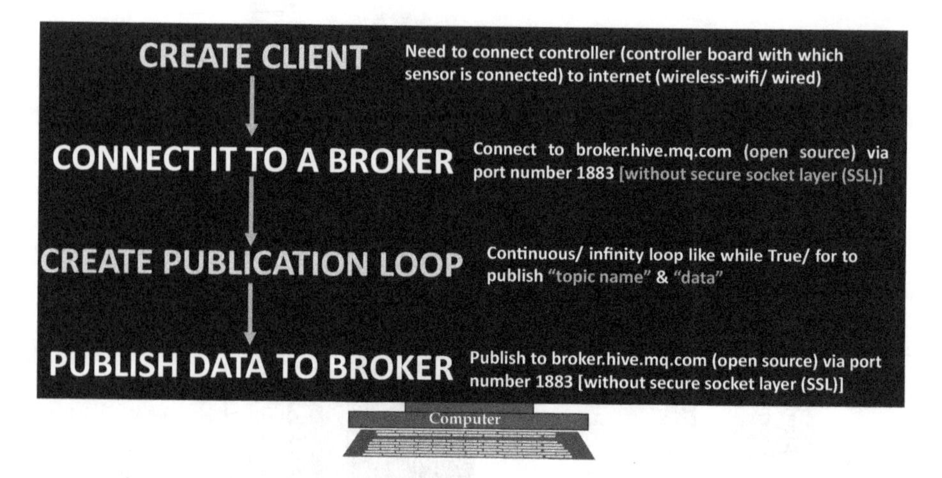

FIGURE 12.3 Program architecture for Rpi computer to communicate the sensor data to broker.

12.5 STEPS TO WRITE A PROGRAM IN RPI COMPUTER

- Connect your Rpi to desktop screen/projector (you can even connect it to your laptop also).
- Provide a power supply to Pi computer.
- Open Thorny Python (Rpi python) is Thonny Python (auto-installed in it). To install python libraries in Rpi, type cmd in search space and press enter.

12.5.1 Working with Microsoft Azure

- Write program as per the above architecture (Figure 12.4) in thorny python.
- Open azure portal. Sign in your Azure portal account, and create IoT hub in Azure services. Further create a device with unique id in IoT hub, and then generate SAS token with shared access policies.
- Figure 12.5 shows the creation of IoT hub on Azure portal, and Figure 12.6 shows the step to create IoT device in IoT hub on Azure portal.
- If required, data can be viewed/displayed in Tableau or Power BI or some other source by calling IoT hub on the platform.
- Create Azure Data Bricks Service in Azure services for data analysis.
- As shown in Figures 12.7 and 12.8, create a new cluster in Azure Data Bricks services for:
  Data injection
  Data distribution
  Cluster management
- As shown in Figure 12.9, create a notebook in the cluster for data injection. Here, we can publish data in IOT hub and connect IOT hub to data bricks
- Fetch data using SQL queries using pi spark in the notebook.
- Write python codes for data analysis.

```
import paho.mqtt.client as mqtt
import ssl
path_to_root_cert = r"D:\Azure IoT\abcd.cer"
device_id = "mydevice"
sas_token ="SharedAccessSignature sr=mybmlhub.a
           zure-devices.net&sig=i6zGMD4%2B7RI
           wF1BlfHg%2FaWBz4OnZfcEqw%2FzBq
           u9vk5E%3D&se=1614"
iot_hub_name = "mybmlhub"
```

FIGURE 12.4 Screenshot of Rpi computer interface to write python codes.

- We can visualize the data.
- Perform various operations to convert unorganized data into organized data.
- On this organized data, we can then perform ML with AI tools.

12.6 APPLICATION AREAS OF AIOT

IoT typically need to have a sensor (with microcontroller), cloud to store data, and internet to connect the two. There are several applications of IoT integrated with ML and AI across multiple industries. Some of these are given as follows:

- **Automated Cleaner**: Robotic vacuum cleaner with a set of sensors to clean dirty spots on the floor with capability to identify obstacles, steep drops etc.
- **Collaborative Robots or Cobots**: Machines designed to help humans from home to office to industry to every phase of day-to-day working and life.
- **Digital Twins**: Digital replica of the real-world object to analyze object's performance without traditional testing methods to testing costs.
- **Drones**: Aircraft without a human pilot.
- **Emotional Analysis**
- **Face Recognition**
- **Security and Access Devices**: Locking/unlocking doors provide access to different machines/equipment etc.
- **Self-Driven Cars**: Self-driving of car is only possible with the integration of AI and IoT.
- **Smart Cities**: The concept of smart cities is conceived with a network of sensors with physical infrastructure of the city for its monitoring, e.g., pollution control, energy efficiency, water usage, traffic monitoring and control, crime control, etc.

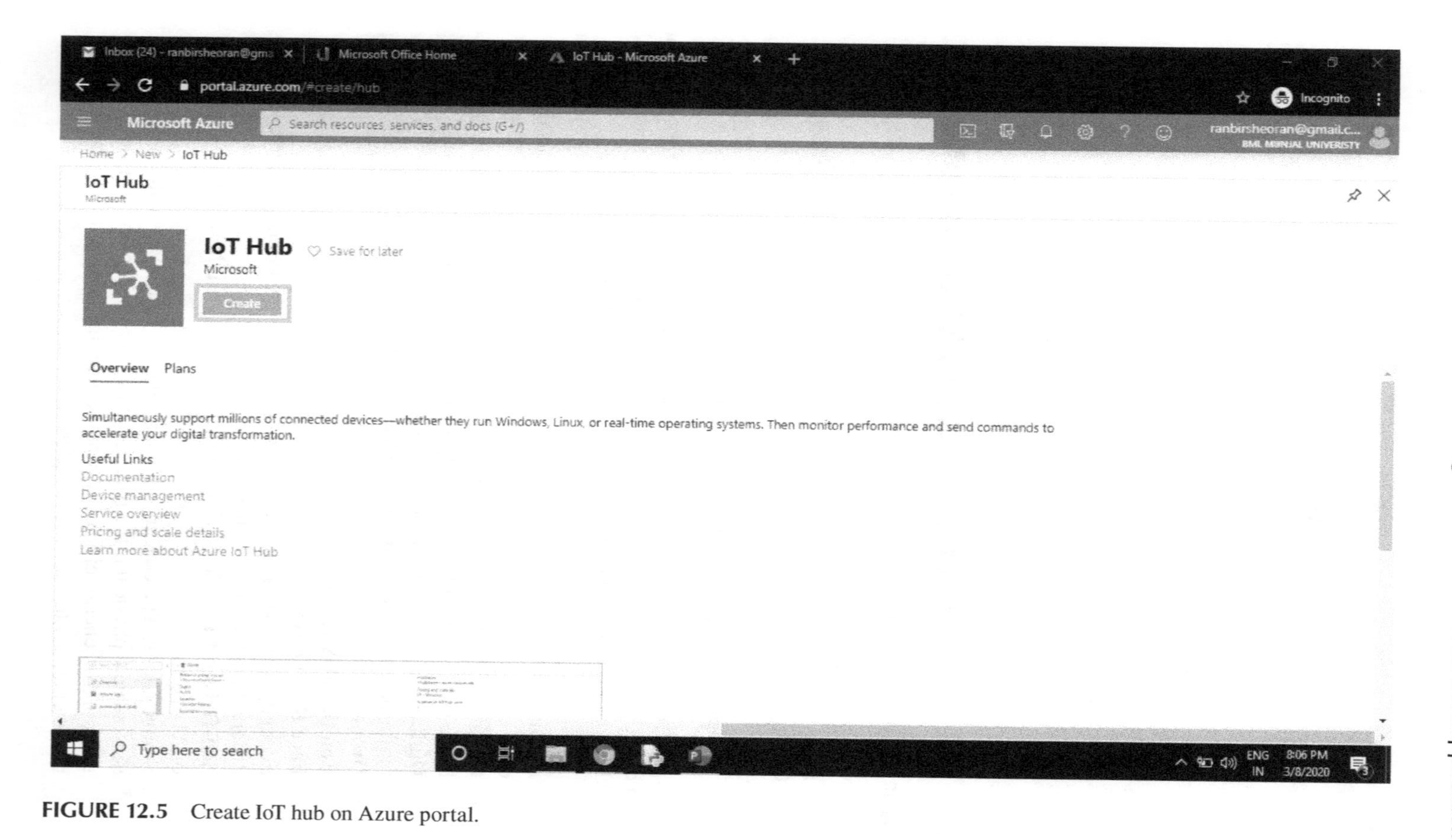

FIGURE 12.5 Create IoT hub on Azure portal.

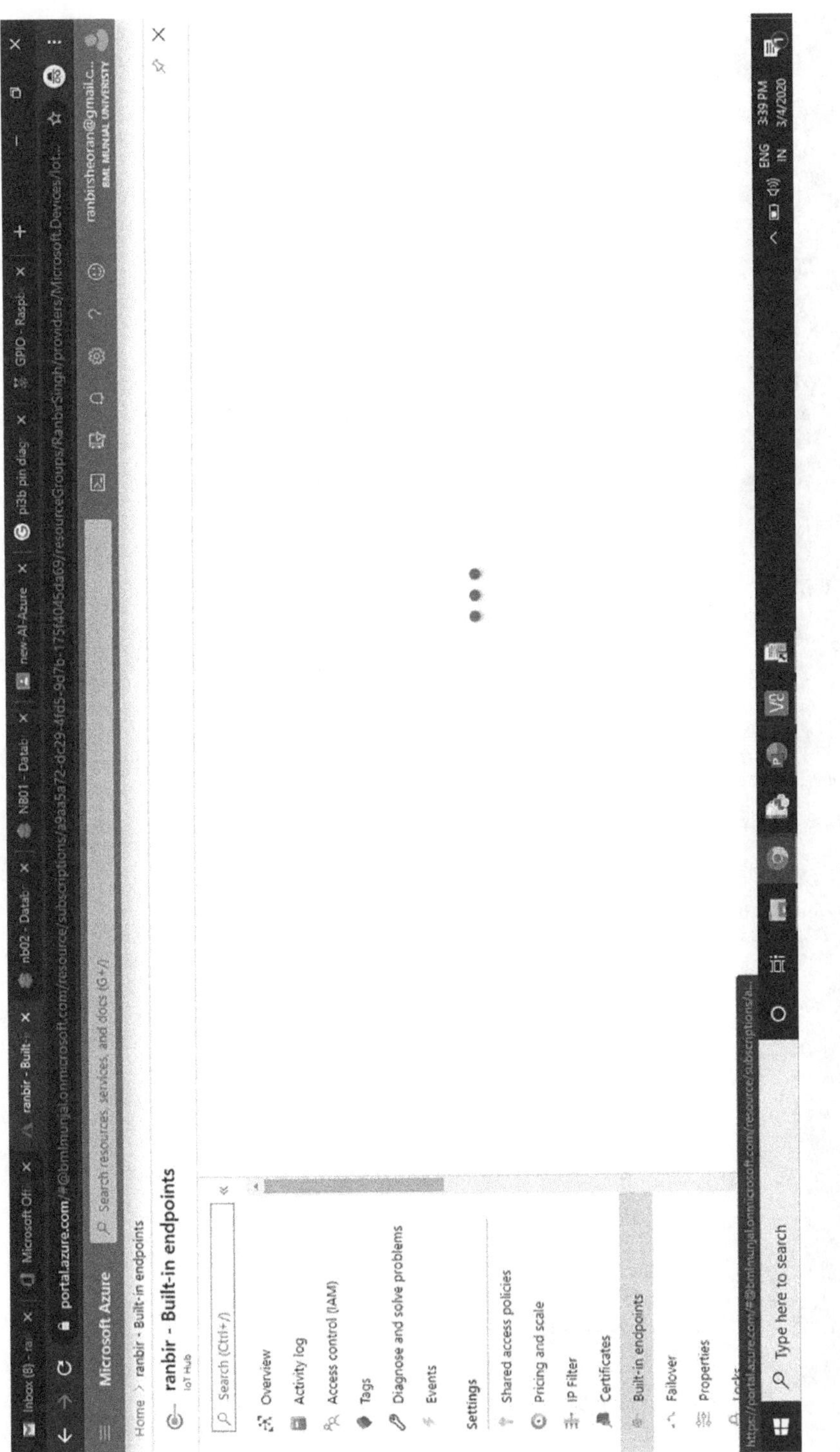

FIGURE 12.6 Create IoT device in IoT hub on Azure portal.

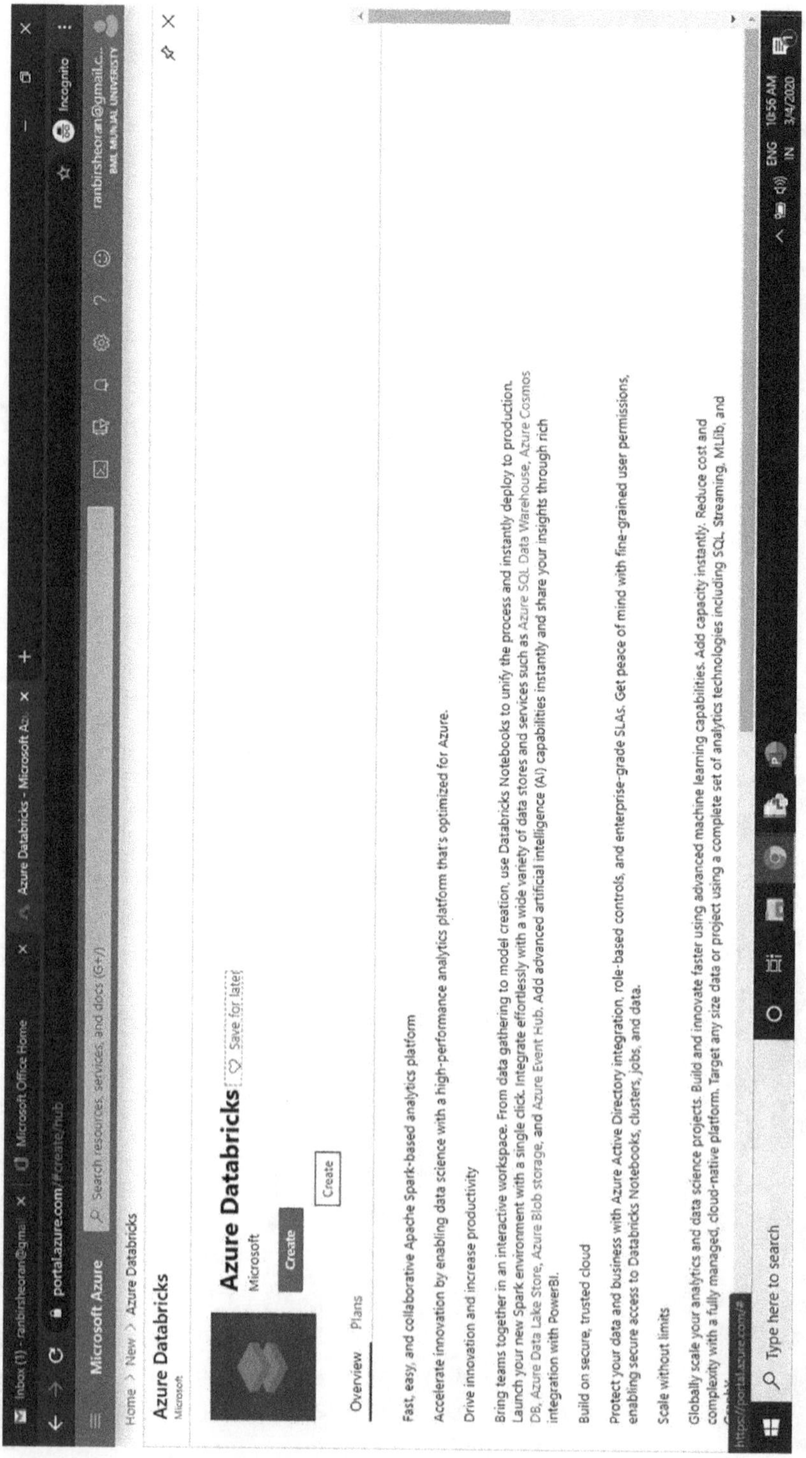

Inbox (1) - ranbirsheoran@gmai
Microsoft Office Home
Azure Databricks - Microsoft Az
portal.azure.com/#create/hub
Incognito
Microsoft Azure
Search resources, services, and docs (G+/)
ranbirsheoran@gmail.c...
BML MUNJAL UNIVERISTY
Home > New > Azure Databricks
Azure Databricks
Microsoft
Azure Databricks
Save for later
Microsoft
Create
Create
Overview
Plans

Fast, easy, and collaborative Apache Spark-based analytics platform

Accelerate innovation by enabling data science with a high-performance analytics platform that's optimized for Azure.

Drive innovation and increase productivity

Bring teams together in an interactive workspace. From data gathering to model creation, use Databricks Notebooks to unify the process and instantly deploy to production. Launch your new Spark environment with a single click. Integrate effortlessly with a wide variety of data stores and services such as Azure SQL Data Warehouse, Azure Cosmos DB, Azure Data Lake Store, Azure Blob storage, and Azure Event Hub. Add advanced artificial intelligence (AI) capabilities instantly and share your insights through rich integration with PowerBI.

Build on secure, trusted cloud

Protect your data and business with Azure Active Directory integration, role-based controls, and enterprise-grade SLAs. Get peace of mind with fine-grained user permissions, enabling secure access to Databricks Notebooks, clusters, jobs, and data.

Scale without limits

Globally scale your analytics and data science projects. Build and innovate faster using advanced machine learning capabilities. Add capacity instantly. Reduce cost and complexity with a fully managed, cloud-native platform. Target any size data or project using a complete set of analytics technologies including SQL, Streaming, MLlib, and

https://portal.azure.com/#
Type here to search
ENG IN
10:56 AM 3/4/2020

FIGURE 12.7 Create Azure Data Bricks on Azure portal.

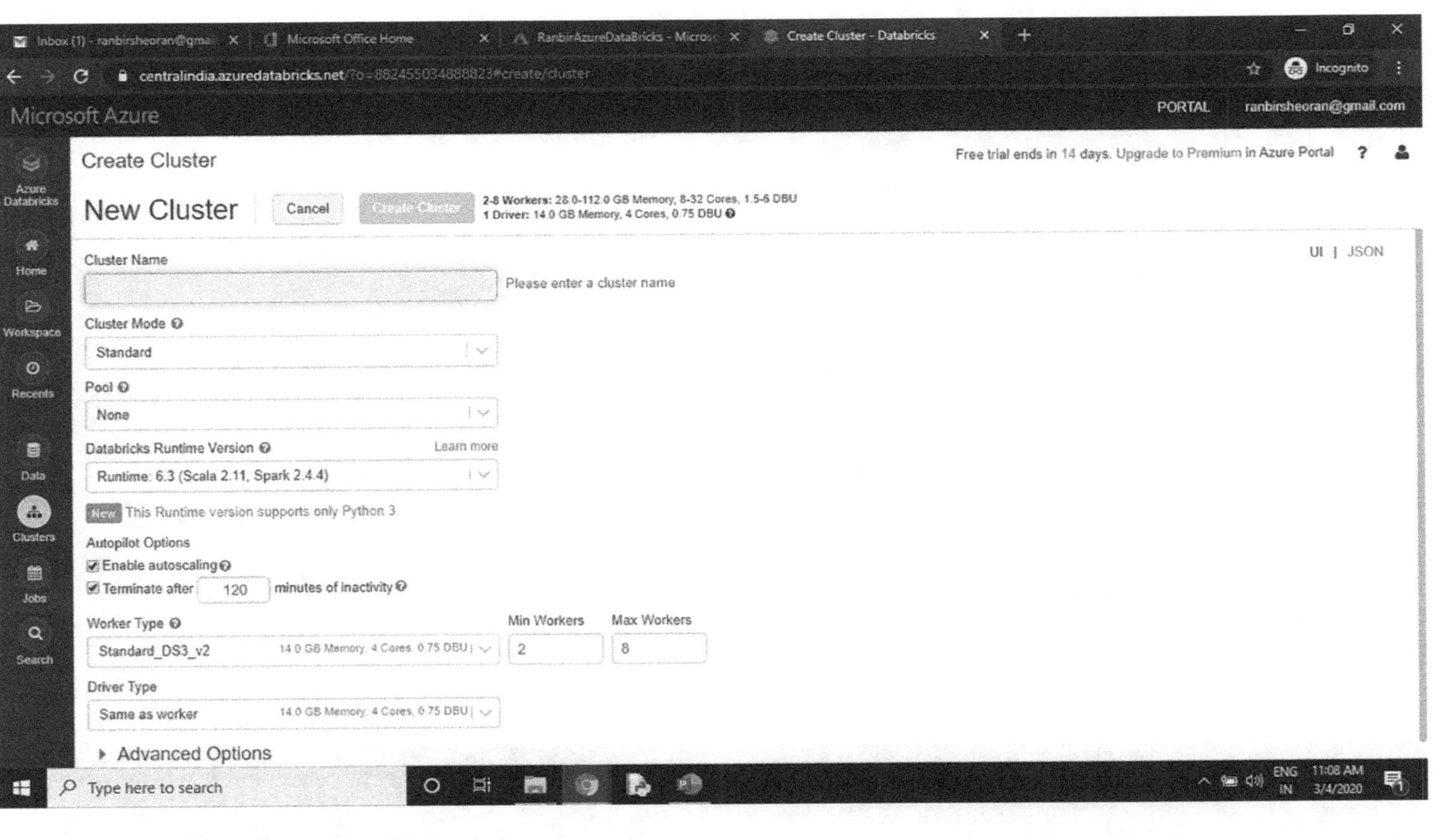

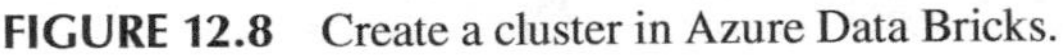
FIGURE 12.8 Create a cluster in Azure Data Bricks.

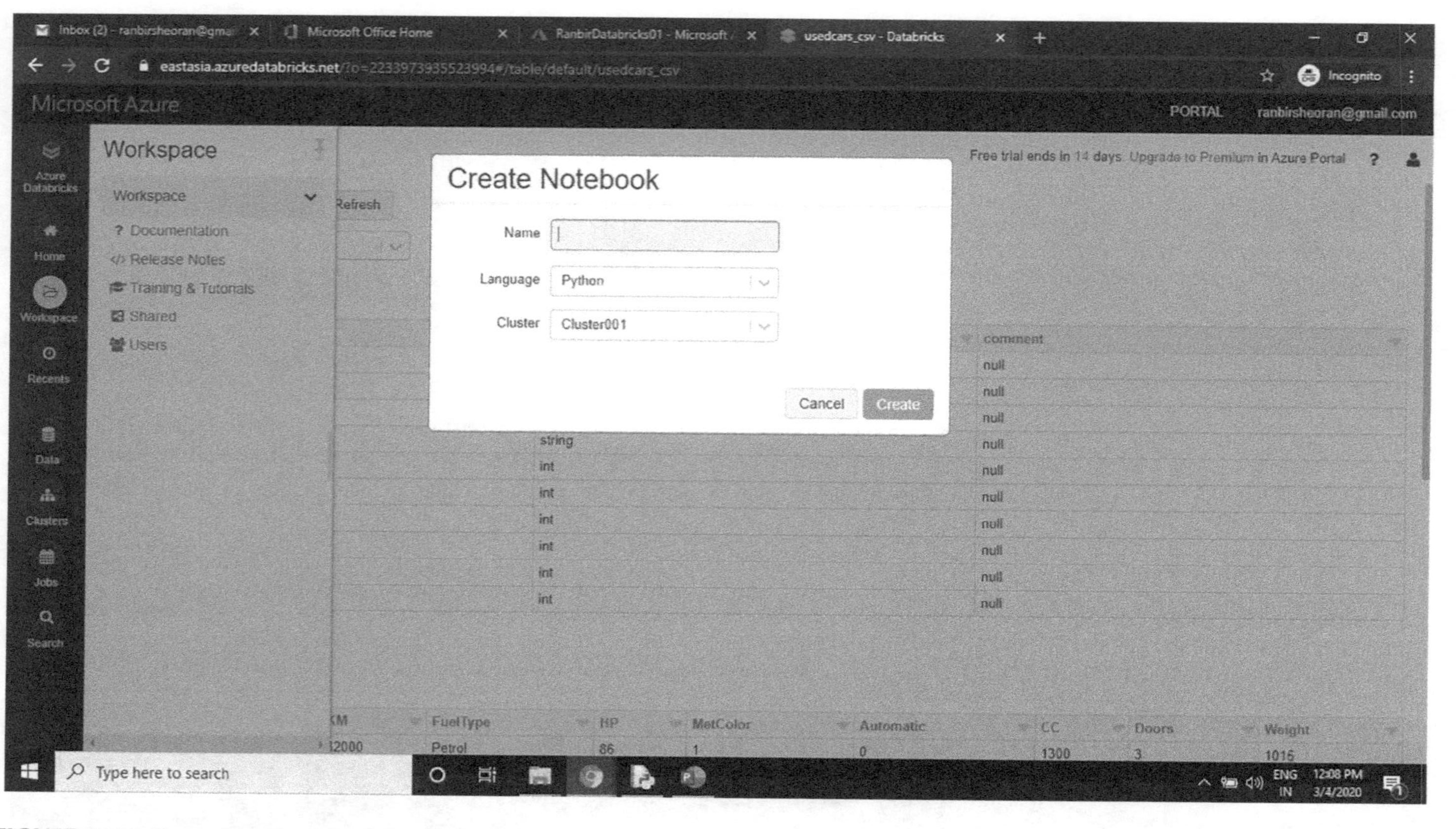

FIGURE 12.9 Create a notebook in cluster in Azure Data Bricks.

- **Smart Homes**: It makes life simpler and easier.
- **Smart Retailing**: Shopping is made smart with the application of AI and IoT to understand customers' behavior.
- **Smart Thermostat**: It allows monitoring and control of temperature from anywhere by its integration with smartphone, IoT device.
- **WildTrack**: IoT and AI algorithms enabled the footprint identification technique (FIT) to identify the species and their age and gender with its unique footprint to recognize their movements, health etc. and to monitor the species.

For these vast range of applications, vertical computing is insufficient in the majority of applications. For the purpose, we know parallel computing, which as discussed is provided by Microsoft as AZURE, Amazon as AWS, and IBM. This chapter is written to work with Azure cloud platform.

12.7 CONCLUSION

IoT-powered AI is a realistic future, which will support, monitor, and control our life. No longer after a decade, you will not be worried about health reminders and issues of your near and dear ones. AI will support you in the task. No longer you will be required to worry about any monitoring and control tasks. Cloud services are attempting to bring all required tools on a single platform to support the technical and business needs of ours.

REFERENCES

1. Bernard Marr, "What is the artificial intelligence of things? When AI meets IoT", *Forbes*, Dec 20, 2019 (https://www.forbes.com/sites/bernardmarr/2019/12/20/what-is-the-artificial-intelligence-of-things-when-ai-meets-iot/#2e2c8f39b1fd).
2. Shanika Perera, "The power of combining AI and IoT", *towards data science*, Oct 2, 2019 (https://towardsdatascience.com/the-power-of-combining-ai-and-iot-4db98ac9f252?gi=8abcad1babeb).
3. Alicia Phaneuf, "Top IoT business opportunities, benefits, and uses in 2020", *Business Insider*, Feb 19, 2020 (https://www.businessinsider.com/iot-business-opportunities-models?IR=T).
4. Mark Patel, Jason Shangkuan & Christopher Thomas, "Adoption of the Internet of Things is proceeding more slowly than expected, but semiconductor companies can help accelerate growth through new technologies and business models", *McKinsey*, May 10, 2017 (https://www.mckinsey.com/industries/semiconductors/our-insights/whats-new-with-the-internet-of-things).
5. Ashish Ghosh, Debasrita Chakraborty & Anwesha Law, "Artificial intelligence in Internet of Things", *CAAI Transactions on Intelligence Technology*, 3, 208–218, 2018.
6. Global artificial intelligence of things (AIoT) solutions market report 2018, Dublin, Feb 12, 2019 (https://PRNewswire.com).

13 Load Balancing in Wireless Heterogeneous Network with Artificial Intelligence

Tanu Kaistha
I.K.G Punjab Technical University

Kiran Ahuja
DAV IET

CONTENTS

13.1 INTRODUCTION

Artificial intelligence (AI) is a system for solving complex problems with a system or machine, and has taken steps without human intervention by witnessing complex natural processes such as learning, reasoning, and self-correction. In computer science, AI, called machine intelligence, is mechanical, unlike the human, intelligence shown by humans. In other words, it is also used to describe machines (or computers) that mimic human-appealing tasks associated with the human mind, such as "learning" and "problem-solving" [1].

Nowadays, AI has become useful in everyday life and has greatly influenced our way of life despite employing basic computer workers, and is not limited to just a PC or a design industry and used in business, medical, legal, educational, manufacturing, and wireless communications [2]. Some examples of AI techniques that are used to date now are spam filters, smart email input, voice elements, self-help text, good personal assistants, automated process, online customers, security monitoring, sales, and business forecasting.

Implementation of AI includes specialist systems, speech recognition, and machine learning (ML) methods, fuzzy logic, deep learning (DL), neural network, and deep reinforcing learning (DRL) less in wireless systems for various purposes that greatly improve the performance of heterogeneous systems and users. It is expected to be used in a variety of fields to conduct the most advanced research in order to improve the efficiency of the short-term data other than manual operations with standard functions such as "identification," forecasting," and "optimization" [3].

QoE and QoS are the two important features of a trusted network, meaning that with any type of implementation, the system is reliable. QoS produces network tests with measurement parameters such as power, bandwidth, error rates, and latency. It is very important for internet services, multimedia apps, online gambling, and different types of video-conferencing, etc., which are based on a consistent, stable, and fast communication. QoE is a user-created feature of a given object. This is the difference between getting the service real and accessing customer service experiences. Reliability is a performance measurement parameter that provides system robustness. The power of an application consistently is called a reliable system. The system is called extremely reliable when it produces the same results in the same environment. Various types of AI help AI systems work with different capabilities.

13.2 DIFFERENT TYPES OF ARTIFICIAL INTELLIGENCE

Nowadays, an AI system can handle amounts of data and make calculations faster. In the coming years, it will reach and surpass human performance in solving different tasks in less time [1,2,10]. In this chapter, we will describe different types of AI. AI can be categorized into different types, and there are two main types of key classifications based on capabilities and based on the AI functionality, shown in Figure 13.1.

13.2.1 Reactive Machines AI

Active AI systems have limited capabilities and can mimic a person's ability to respond with different types of incentives. These systems cannot use previous observations to

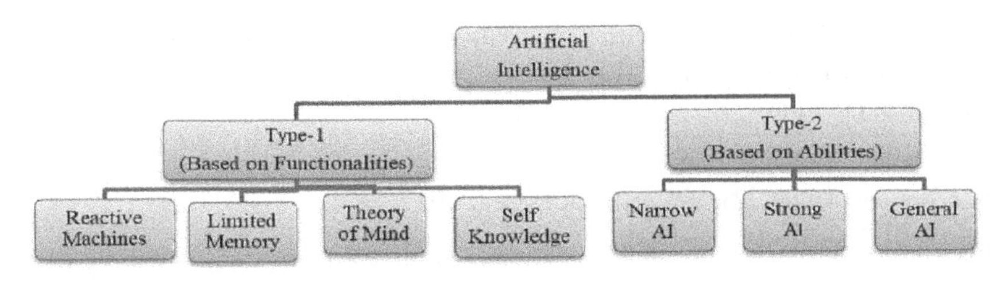

FIGURE 13.1 Types of artificial intelligence.

inform their current actions, which means they have no memory-based functionality; e.g., these machines cannot "learn." They cannot be used to rely on memory to improve their performance based on similarities and also not respond automatically to the limited input, for example, IBM's Deep Blue.

13.2.2 Limited Memory AI

These memory machines are powerful and unique for reading historical data to make decisions. AI systems, such as DL, have been trained with large amounts of data training that they store in their memory to solve future problems and increase accuracy. For example, virtual assistants for self-driving cars and holding phones are all driven by the limited AI memory.

13.2.3 Theory of Mind AI

This is the newest technology, and it will be able to understand the organizations they work with by understanding their beliefs, needs, feelings, and processes of thinking. For AI researchers, accessing theory level requires the development of other branches of AI and understands human necessities whose minds can be shaped by different factors.

13.2.4 Self-Knowledge AI

This is very similar to the human brain until it is created. This will not only be able to understand and evoke emotions in those interactions, but also should have feelings, needs, and beliefs, and may wish for it as AI would be able to have ideas such as self-monitoring.

13.2.5 Artificial Narrow Intelligence (ANI)

This system can only perform specific tasks independently using human-like capabilities. These machines can do what they can do, and thus have very limited or little distance skills that all have a quick and limited AI memory. The most sophisticated AI uses ML and DL teaching methods that themselves fall under the ANI.

13.2.6 Artificial General Intelligence (AGI)

This is the ability to learn and understand a job like absolutely a human being. These programs will be able to independently build more capacity, connectivity, and logistics, and often spread to different locations during the time required for training programs. This will make AI systems as powerful as humans by multiplying our multitask capabilities.

13.2.7 Artificial Strong Intelligence (ASI)

This kind of AI has super intelligence: you can do for us or think anything like a human. The robot "Alpha 2" is an example of this category, which has the competency to manage a smart home; predict the weather; and entertain you with exciting news, games, and music in the most effective way.

The different types of AI are selected, according to their advantages and appropriateness to the system.

13.3 ADVANTAGES OF ARTIFICIAL INTELLIGENCE

Nowadays, AI is becoming the most popular quality of most applications due to its number of benefits, and few of them are discussed below.

Low Error and Security: AI is carefully customized, at which point the chances of system errors are low compared to humans. Fairness, clarity, truth, and speed are particularly high and are not influenced by hostile conditions. Circumstances could be detrimental to human-like exploration; high-temperature mines care less about the AI system. The best robots in this regard can create dangerous bonds, can constantly think in unforgivable situations, and can make good decisions with little or no mistakes. The AI system acts as a security agent for our homes, cars, and organizations, and helps manage records in the safest way possible.

Easy Work Capability and Effortless: It is a multifaceted, meticulous workout in many places carefully without any human help. It also simplifies the work in mining and the power of burrowing, and can detect work conditions before they start. Construction engineering programs run as a trained participant in basic training needs and do not require sleep, rest, breaks, or relaxation. These systems do not get fed up or tired.

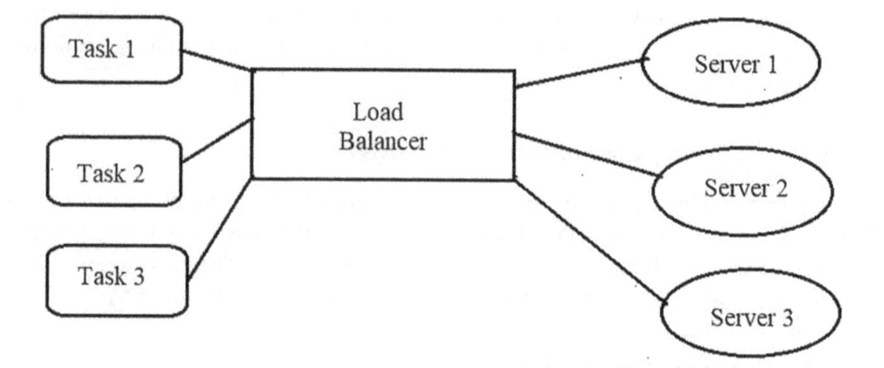

FIGURE 13.2 Basic load balancer.

Medical Help: It can be used effectively for medical purposes (e.g., health risks and interest status) and can provide a medical response data. Radios and various forms of the medical process can, in time, achieve more man-made accuracy.

13.4 DISADVANTAGES OF ARTIFICIAL INTELLIGENCE

While actualizing in the systems, the following are the disadvantages of AI, which can't be ignored.

Cost: AI systems require huge amounts of time and money to build, rebuild, and maintain. For example, robot costs more and has more resources than any program.

Understanding and Storage: AI is expandable, depending on the opportunity and regaining data in memory as humans; whenever properly adapted, they can understand and improve on assignments in any situation. Machine intelligence and human thought can never be integrated with AI.

Misuse: Smart phone abuse and other technologies have made people rely on AI and decreased their understanding of fitness. Devices will undoubtedly be used to destroy, whenever they put the wrong hand that would be afraid of mankind.

In addition to the advantages and disadvantages, the system acquires AI as indicated by its requirements/uses. AI techniques make the process extremely elegant as indicated by its requirements.

13.5 ARTIFICIAL INTELLIGENCE: METHODS AND APPLICATIONS

There are the following methods used for AI [8,9,10].

Machine Learning (ML): It is an investigation of computer algorithms that develop appropriately for understanding and further establishing AI. These types of algorithms are a numerical model that relies on the sample data, known as "training data," to be stable in prediction or selectivity without explicitly programmed to do it and have applications, for example, e-mail distinction and PC vision.

Genetic Algorithm (GA): Genetic algorithms (GAs) are used to produce high-throughput algorithms and search problems based on naturally occurring functions, for example, transformation, crossover, and determination.

Neural Networks (NN): This is the set of algorithms, freely displayed in the back of the human brain, intended to detect the formation. These algorithms translate sensory information through a sort of machine observation and marking.

Fuzzy Control System (FCS): The word "fuzzy" means to the fact that the logic involved can manage with ideas that can't be communicated as the "true" or "false" but rather as "partially true." It has the potential that the solution to a problem can be thrown out in terms that humans can understand, so that their information can be used in the design of a controller that facilitates the tasks that have been successfully performed by humans.

Evolutionary Algorithm (EA): In AI, an evolutionary algorithm (EA) is a widely used set of algorithms, a common generalization algorithm for humans. This uses the methods inspired by biological evolution, such as reproduction, genetic modification and regeneration, and selection.

Hidden Markov Model (HMM): Hidden Markov model (HMM) is a statistical Markov model in which the system being modeled is assumed to be a Markov process with unobservable ("hidden") states and only observations can be seen. Such a process can be trained for several applications such as speech recognition and computational biology.

AI is utilized in various types of fields like science, medical, communication, and business, and is assuming a significant job in our everyday life and future. AI is a feasible solution for the emerging complex communication system design. The rise of AI technology can provide a more intelligent and effective strategy for achieving a wireless communication to enhance the use of range. In the following point, we will talk about the significance of AI in wireless heterogeneous systems. However, there have been many suggested ways to improve the effectiveness of heterogeneous network-proposed techniques for directly improving HetNets performance. In the next point, we will talk about the significance of AI in wireless heterogeneous systems.

13.6 AI IN WIRELESS HETEROGENEOUS NETWORKS (WHN)

Nowadays, due to the rapid development of the mobile telecommunications industry and the internet, the traffic load is seen by explosive growth over the last and according to optimized mobile applications and improved wire access with a variety of skills, problems, functionalities to the heterogeneous network. In a heterogeneous network, there are distinct types of cells such as macrocell, base station (BS), and NodeB (eNBs) developed in the long-term evolution (LTE), picocell and femtocell cells with different types of transmission, cover, and operating systems. The principle target of heterogeneous systems is to expand limits, load balancing, change cell edge coverage area, and effective utilization of range [12]. The biggest issue for a high-power system is the development of productive techniques for the formal distribution of loads between different loads with different load balancing techniques [5]. Due to the functionality of the heterogeneous network, performance should continue to make better for a number of features, such as power conversion, load balancing, mobility management, flexibility, neighbor optimization, and more user needs in the workplace.

In the next topic, we discuss the importance of measuring load in the AI system with different techniques.

13.7 IMPORTANCE OF LOAD BALANCING IN AI

Load balancing is the process of allocating all workloads to all distributed system domains to optimize usage and resource utilization; analyze bandwidth and response time; and avoid overload and under-load conditions. Figure 13.2 shows the basic structure of the load balancing concept. It is needed to get the best execution of each small header in different processes [5].

Although there are many suggested ways to improve the execution of heterogeneous network, it has attracted a great interest from industry and academia by implementing AI techniques, e.g., neural network, ML, and so on. AI techniques can automatically solve problems for wide systems, such as a heterogeneous network with those techniques that are intelligent and automated. The advanced features in self-organizing networks (SONs) can dramatically reduce human engagement in workflows, coverage, and QoS, and can increase the network capacity in the heterogeneous network [14,15]. The different load balancing techniques in AI to get an efficient network are discussed in Refs. [16,17,18].

13.6.1 Machine Learning in a Wireless Heterogeneous Network

ML emerged from the pattern recognition studies and computational learning in AI environments, and studied the human brain [19]. It explored the design and study of algorithms, and made algorithms to read and adapt to changing environments in the wireless network [22]. It was described by a machine-learning method called Q-learning to detect a supply chain problem on parallel networks [23,25]. It makes the system fit to consequently take in and improve from its encounters, and starts with perceptions or information. It creates a scientific model of the first-hand data, known as "training data," to solve forecast or selections and enable the ability to investigate the wide amounts of data with immediate and consistent results. The collaboration of ML with AI and the advances in use can make it compelling to investigate large amounts of data. There are various methods of ML, such as supervised, unsupervised, semi-supervised, and reinforcement learning (RL). Other benefits of ML reduce over-signing, provide better performance than traditional ones, avoid previous errors, improve algorithm complexity, compute power, and overcome network information shortages [26].

13.6.2 Neural Network in a Wireless Heterogeneous Network

The Artificial Neural Network (ANN) is a model for preparing data that is inspired by biological sensors, which is similar to the details of the brain process, and contains a very large amount of neurons to detect specific issues. It consists of numerous nodes, which mimic the neurological features of the human brain connected to the joints that have been evaluated by weight to communicate between them. The final output of each node is called its node value, when it receives the input information and performs the function; the result is sent to the next neurons. ANNs can learn by changing weight esteems. There are two topologies of the ANN top, namely, the feed and the feedback. A trained neural system can go "expert" in the classification of data that will provide bearings in unpredictable situations. An ANN-based handoff decision algorithm is used to reduce the inactivity of the wireless infrastructure [4]. There are other benefits of the neural network such as information capacity, readability, and real-time power, and can be stored if neurons fail.

13.6.3 Fuzzy Logic for a Wireless Network

Fuzzy logic is a logic system that has more values based on "degrees of truth" than the usual "true or false" (1 or 0) Boolean logic on which a modern computer is based. It has been used to capture the concept of an empirical fact, where the true value may differ between extreme and absolute truth. It includes 0 and 1 as cases of extreme truth (or "state of things" or "truth") but also includes various regions of truth between fuzzy logic developed in various areas, ranging from advanced logic to AI. In fuzzification, the mathematical input values map into fuzzy membership functions [6,21]. In contrast, the de-fuzzification functions can be used to map the output of a dense output to a "crisp" output value that may be used for decision or control purposes.

An optimized logic algorithm has been proposed to improve the transmission quality with a minimal management delay, minimal packet loss, and poor handling management. Fuzzy logic applications have been proposed to be measured by channel, channel size, and channel layout on mobile networks [7,20,24]. The main benefits of a practical idea are simple and design; it can solve the problem of impartial installation, validation data, variations in the languages used, and a relatively stable system with a limited amount of time.

13.6.4 Genetic Algorithm

A GA was developed to provide the optimum utilization of resources in a wireless environment, and also confirms the QoS requirement for customer service. In this case, variations in the ability to manipulate and select can be used as a future activity for further good results and presentations. Among the many EAs, GA must be discovered and designed to perform complex systems [6]. GA adopts the process of genetic modification, which involves the two key functions, the crossover, which makes easier discovery of the correct solution, and the transformation of genes. In the most open cases, the GA is skilled at solving problems where the size of the solution is too high to be fully searchable, and can be properly converted to one (or more) correct results, in a short period of time and can solve many multipurpose problems easily. Therefore, GAs are widely used to develop heterogeneous networks, especially in cell organization and spatial induction where a large number of parameters need to be evaluated [11,17]. For example, a multipurpose GA addresses the multipurpose communication problem of the heterogeneous network placement problem, aiming to increase the communication coverage and total bandwidth, and to reduce cost savings.

13.6.5 Particle Swarm Optimization (PSO)

Particle swarm optimization (PSO) is a high-quality, population-based, and global algorithm based on swarm intelligence, which finds a solution from the problem of optimizing search space and social behavior. A social network is defined and assigns to each person to participate; then a population that is defined as a random guess on a set of problems is introduced and is referred to as the solution of the candidate and also known as particles. The particles calculate the robustness of the candidate solutions and remember the area where they were most successful [17]. The best solution for a person is called the best particle of space and gets the information from its neighbors. The position of the particles is influenced by the best position visited personally, i.e., its experience and position of the best particles in its area. When the aggregate particle neighbor is violent, the best position in the neighborhood is called the global best particle, and the resulting algorithm is called PSO gbest. When small neighborhoods are used, the algorithm is often referred to as the lbest PSO. Each particle's performance is measured using a density function that varies depending on the accessibility problem.

13.6.6 Artificial Bee Colony (ABC)

The artificial bee colony (ABC) algorithm was introduced in 2005, and it is the meta-heuristic algorithm used. This algorithm mimics the process of making honey bees and has three stages. There are employee bees, onlooker bees, and scout bees. In the employee bee and onlooker bee categories, bees create local researchers' sources of available neighborhoods based on selections prepared in the bees section employed and probabilistic selection in the onlooker bee category [13]. In the scout bee category, which symbolizes the abandonment of energy-consuming food resources in the process, the solutions no longer work when they seek to find new regions in space and have good exploration and developmental potential.

13.6.7 Markov Models and Bayesian-Based Games

The Markov models and Bayesian methods are not AI methods, but still provide mathematical solutions for the particular networks with the ability to transform them [28]. As well as the changing dynamics of the state between the different regions in both learning models, home and navigation systems on the higher-level networks can be improved based on the tree depletion. Achieving the right types of local HMMs is often used to adjust the signal strength history. This study builds on the problem of network selection in a heterogeneous network with incomplete data by combining large reaction energies. It is also investigating the use of specific Bayesian methods for anomaly detection in a heterogeneous network, which can provide human interpretations.

The methods discussed above are used to handle various aspects of wireless communication such as network management, mobility management, traffic loading, load loading, etc. in the heterogeneous network.

13.8 CONCLUSION

Unless there are multiple algorithms for finding load balancing in a heterogeneous wireless system, the problem is to efficiently manage the multidimensional nature of the heterogeneous systems using transformational strategies. The heterogeneous network will be characterized by a high degree of capillarity, population, and high bitrate. Any process selected depends on the different parameters, the tradeoff between these components, the application, the multimode gadget, and more. By integrating AI-based SON features, a heterogeneous network may have a great capability to demonstrate the potential of mobile service and be the best way of doing things in measuring the network load. Reliability refers to the power of the system regularly, while the practical steps take the system out of a given time frame. Few mistakes lead to greater performance over time. Most of the offerings come from replica adapter, excellent automation capabilities for system development, and application of dynamic processes, as well as a fault tolerance system functionality. Overall by this research and analysts, AI-based strategies are explored and demonstrated the ability to develop and master the best development of the different load balancing techniques in the heterogeneous network and its technical challenges and research problems.

REFERENCES

1. Charniak E. *Introduction to Artificial Intelligence*. Pearson Education India, Noida; 1985.
2. Nilsson NJ, Nilsson NJ. *Artificial Intelligence: A New Synthesis*. Morgan Kaufmann, San Francisco, CA; 1998 Apr 15.
3. Russell S, Norvig P. *Artificial Intelligence: A Modern Approach*. Prentice Hall/Pearson Education, Denver, CO; 2002.
4. Nasser N, Guizani S, Al-Masri E. Middleware vertical handoff manager: A neural network-based solution. In *2007 IEEE International Conference on Communications*, 2007 Jun 24 (pp. 5671–5676). IEEE, Glasgow.
5. Lee M, Ye X, Marconett D, Johnson S, Vemuri R, Yoo SB. Autonomous network management using cooperative learning for network-wide load balancing in heterogeneous networks. In *IEEE GLOBECOM 2008–2008 IEEE Global Telecommunications Conference*, 2008 Nov (pp. 1–5). IEEE, New Orleans, LA.
6. Alkhawlani M, Ayesh A. Access network selection based on fuzzy logic and genetic algorithms. *Advances in Artificial Intelligence*. 2008;2008:105.
7. Shi WX, Fan SS, Wang N, Xia CJ. Fuzzy neural network based access selection algorithm in heterogeneous wireless networks. *Journal of China Institute of Communications*. 2010 Sep;31(9):151–6.
8. Demestichas P, Georgakopoulos A, Karvounas D, Tsagkaris K, Stavroulaki V, Lu J, Xiong C, Yao J. 5G on the horizon: Key challenges for the radio-access network. *IEEE Vehicular Technology Magazine*. 2013 Jul 25;8(3):47–53.
9. Muñoz P, Barco R, de la Bandera I. Optimization of load balancing using fuzzy Q-learning for next generation wireless networks. *Expert Systems with Applications*. 2013 Mar 1;40(4):984–94.
10. Mitchell RS, Michalski JG, Carbonell TM. *An Artificial Intelligence Approach*. Springer, Berlin; 2013.
11. Dasgupta K, Mandal B, Dutta P, Mandal JK, Dam S. A genetic algorithm (GA) based load balancing strategy for cloud computing. *Procedia Technology*. 2013 Dec;10(2):340–7.
12. Nilsson NJ. *Principles of Artificial Intelligence*. Morgan Kaufmann, San Franciso, CA; 2014 Jun 28.
13. Miku DN, Gulia P. Improve performance of load balancing using artificial bee colony in grid computing. *International Journal of Computer Applications*. 2014 Jan 1;86(14):1–5.
14. El-Zoghdy SF, Ghoniemy S. A survey of load balancing in high-performance distributed computing systems. *International Journal of Advanced Computing Research*. 2014;1–3.
15. Chen X, Wu J, Cai Y, Zhang H, Chen T. Energy-efficiency oriented traffic offloading in wireless networks: A brief survey and a learning approach for heterogeneous cellular networks. *IEEE Journal on Selected Areas in Communications*. 2015 Jan 16;33(4):627–40.
16. Jafari AH, Shahhoseini HS. A reinforcement routing algorithm with access selection in the multi–hop multi–interface networks. *Journal of Electrical Engineering*. 2015 Mar 1;66(2):70–8.
17. Wang X, Li X, Leung VC. Artificial intelligence-based techniques for emerging heterogeneous network: State of the arts, opportunities, and challenges. *IEEE Access*. 2015 Aug 11;3:1379–91.

18. Simsek M, Bennis M, Guvenc I. Mobility management in HetNets: A learning-based perspective. *EURASIP Journal on Wireless Communications and Networking*. 2015 Dec 1;2015(1):26.
19. Jiang C, Zhang H, Ren Y, Han Z, Chen KC, Hanzo L. Machine learning paradigms for next-generation wireless networks. *IEEE Wireless Communications*. 2016 Dec 20; 24(2):98–105.
20. Han Q, Yang B, Chen C, Guan X. Energy-aware and QoS-aware load balancing for HetNets powered by renewable energy. *Computer Networks*. 2016 Jan 15;94:250–62.
21. Prithiviraj A, Krishnamoorthy K, Vinothini B. Fuzzy logic based decision making algorithm to optimize the handover performance in Hetnets. *Circuits and Systems*. 2016;7(11):3756.
22. Chen M, Challita U, Saad W, Yin C, Debbah M. Machine learning for wireless networks with artificial intelligence: A tutorial on neural networks. arXiv preprint arXiv:1710.02913. 2017 Oct 9.
23. Latif S, Pervez F, Usama M, Qadir J. Artificial intelligence as an enabler for cognitive self-organizing future networks. arXiv preprint arXiv:1702.02823. 2017 Feb 9.
24. Nyambati ET, Oduol VK. Analysis of the impact of fuzzy logic algorithm on handover decision in a cellular network. *International Journal for Innovation Education and Research*. 2017;5(5):46–62.
25. Kibria MG, Nguyen K, Villardi GP, Zhao O, Ishizu K, Kojima F. Big data analytics, machine learning, and artificial intelligence in next-generation wireless networks. *IEEE Access*. 2018 May 17;6:32328–38.
26. Amiri R, Mehrpouyan H, Fridman L, Mallik RK, Nallanathan A, Matolak D. A machine learning approach for power allocation in HetNets considering QoS. In *2018 IEEE International Conference on Communications (ICC)*, 2018 May 20 (pp. 1–7). IEEE, Kansas City, MO.
27. Zia K, Javed N, Sial MN, Ahmed S, Iram H, Pirzada AA. A survey of conventional and artificial intelligence/learning based resource allocation and interference mitigation schemes in D2D enabled networks. arXiv preprint arXiv:1809.08748. 2018 Sep 24.
28. Gacanin H. Autonomous wireless systems with artificial intelligence: A knowledge management perspective. *IEEE Vehicular Technology Magazine*. 2019 Jul 11;14(3):51–9.

14 Applications of Artificial Intelligence Techniques in the Power Systems

Rishav Sharma
Lovely Professional University

Cherry Bhargava
Symbiosis Institute of Technology

CONTENTS

14.1 INTRODUCTION

Power system is an interconnection of electrical components that are used for generation, transmission, distribution, and consumption of electric power. Motors, generators, and transformers are the components of the power system (Zhu et al., 2008). There are enormous applications of artificial intelligence (AI) in the era of power electronics. AI is the intelligence system exhibited by software and machine: for example, personal assistants like Siri, Alexa, and robots are used for various applications etc.

The AI is utilized for developing the system with features and attributes like humans, i.e., ability for thinking, reasoning, to find the meaning, to differentiate, learning from the mistake, and also to rectify them. AI is the system having the ability to achieve any assignment completely based on the application (Dekker, 1996).

14.1.1 Need of Artificial Intelligence in Power System

The analysis of the power system by traditional techniques has become hard due to the following reasons:

i. The structure of the power system is very vast, and it is difficult to calculate the different parameters in the power system manually.
ii. Traditional techniques take larger computational time, and also, there is difficulty in the accuracy due to the large power system (data handling).

The today's modern power system works in limits due to continuously increasing power consumption and the expansion of the presently existing power transmission grids and lines. This condition is required because the power system operation and control are accomplished by a regular monitoring of the entire system in a detailed manner. Modern computer technology has different tools for the analysis of the power system, for example, the power system operation, control, planning, and monitoring. Out of these computer tools, the AI technique has been increasing exponentially in the last few years and has been put in different areas of the power system (Diehl, 2003).

14.2 TYPES AND CLASSIFICATION OF ARTIFICIAL INTELLIGENT TECHNIQUES

This section deals with the types and classifications of AI techniques, along with their advantages and disadvantages.

14.2.1 Artificial Neural Network

The biologically inspired artificial neural network (ANN) can change the different inputs into a different output by the neuron network, where every neuron gives an output based on the input. Neuron can also be named as a processor. The analysis of neuron function and its unique structure can help in the synthesis of various functions of computer like pattern recognition and interpretation (Samanta & Al-Balushi, 2003).

Each neuron is attached with other neuron by a link as shown in Figure 14.1.

Every link is related to a weight that gives the data and most valuable information about the neuron to perform different applications. Every neuron has an activation signal in its internal state. The combination of input and activation signal provides the output signal, which is further transferred to other units.

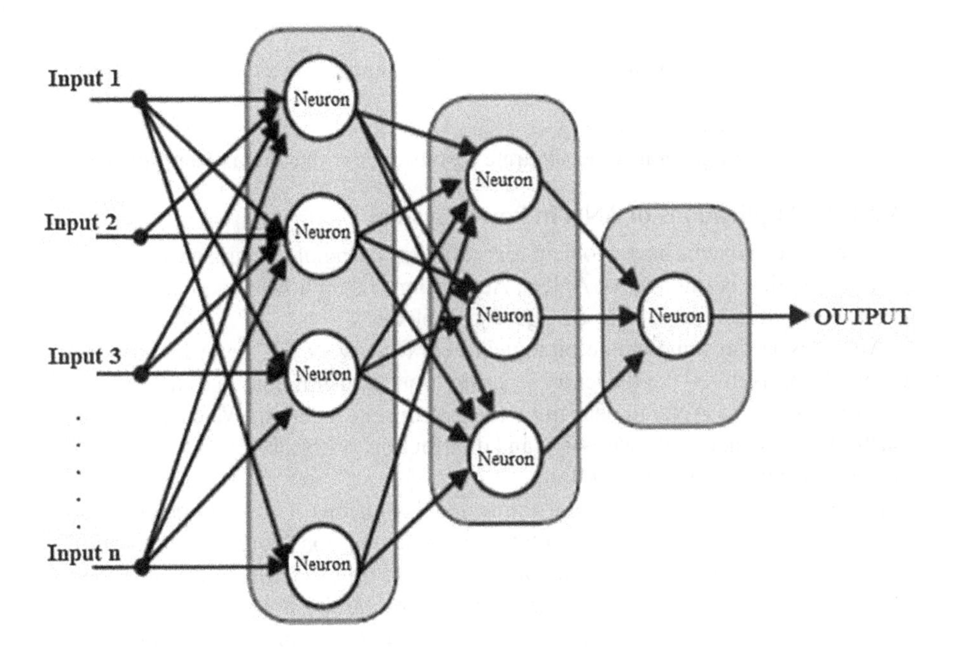

FIGURE 14.1 ANN feedforward network.

14.2.1.1 Classification of Artificial Neural Network

An ANN can be classified by the feedforward or recurrent and connectivity patterns (Stanton, 2009).

Furthermore, the ANN has three layers: input layer, hidden layer, and output layer. The role of each layer is described as follows:

i. **Input Layer**: The nodes of the input layer distribute the data and information to the hidden layer without processing it.
ii. **Hidden Layer**: The nodes of the hidden layer are able to map and solve the nonlinear problems.
iii. **Output Layer**: The nodes of the output layer provide the output to other units.

14.2.1.2 Advantages and Disadvantages of Artificial Neural Network

Advantages:

Using ANN for prediction and estimation purposes has the following advantages and benefits:

i. ANN has a high speed of processing.
ii. ANN has the capability to deal with the situation of insufficient data and information.
iii. ANN is able to tolerate the faults.

Disadvantages:

Using ANN, the following drawbacks or disadvantages often occur:

i. ANN is large dimensionally.
ii. ANN always generates the accurate results even if the data is inappropriate.

14.2.1.3 Applications of ANN in Power System

Modern power systems have problem regarding encoding of an inappropriate nonlinear function that is suitable for ANNs. According to their advantages such as speed, they can give a faster result for emergency problems.

An ANN is the same as the biological neural network for the evaluation of the real-world problems—the difficulty in generation, transmission, and distribution of power is sent to the ANN to determine the appropriate solution. We have to feed the constraints of an actual transmission and distribution system to get an actual value of parameters (Dylis & Priore, 2001).

For example, the parameters of the transmission line, i.e., inductance, capacitance, and resistance, can be calculated by ANN by considering the factors like unbalancing conditions, environment factors, and other factors. Another example of ANN in the power system can be evaluated by taking the parameters of resistance, capacitance, and inductance as an input to ANN and combining with the preset value. The resultant parameter can be able to overcome the proximity and skin effect.

14.2.2 Fuzzy Logic

Fuzzy logic system (FLS) is the type of logical system having different standards and rules to give a definite output as shown in Figure 14.2.

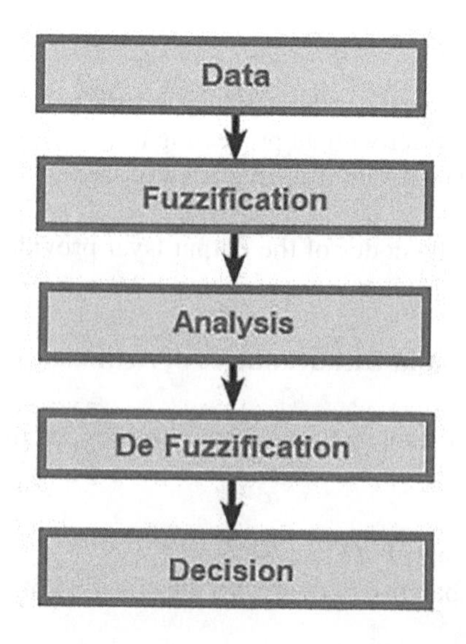

FIGURE 14.2 Flowchart of fuzzy logic system.

FLS is the same as human decision-making system having the capability to give an appropriate solution even for the distorted, incomplete, inappropriate, and inaccurate fuzzy inputs. Fuzzy logic is the system having the same working capacity as the human mind, by taking the advantage of human decision-making in machine so that we can execute or operate the machines as humans (Hayward & Davidson, 2003). The process of converting or changing the simple input into the fuzzified output is known as fuzzification. FLS is simple, accurate, and easy to implement system, and it is also used to execute many applications. In the power system, the FLS is best suitable for application where an immediate decision-making is required.

14.2.2.1 Advantages and Disadvantages of Fuzzy Logic

Advantages:

Using fuzzy logic for estimation and prediction purposes has the following advantages and benefits:

i. FLS is flexible.
ii. FLS is easy to develop and recognize.
iii. FLS has the ability to give the solution for the complex problem.

Disadvantages

Using fuzzy logic, the following drawbacks or disadvantages often occur:

i. FLS is not following the systematic approach for designing the fuzzy logic.
ii. FLS is used only when there is no need of high perfection.

14.2.2.2 Applications of Fuzzy Logic in Power System

It can be used for planning and plotting the various components from small to large, and also improves the efficiency of the various components in the modern power system.

i. FLS used in planning and controlling the power system.
ii. It is used for fault diagnosis.
iii. Load forecasting can be evaluated by the fuzzy logic.

14.2.3 Expert System

Expert system (ES) in the field of AI is one of the important research domains. ES acquires the understanding of humans and utilizes this knowledge into machine execution (Bhargava, Banga, & Singh, 2016). ES is the form of computer program designed to solve different works in a particular field. The block diagram of ES is shown in Figure 14.3.

14.2.3.1 Advantages and Disadvantages of Expert System

Using ES for the estimation and prediction purposes has the following advantages and benefits:

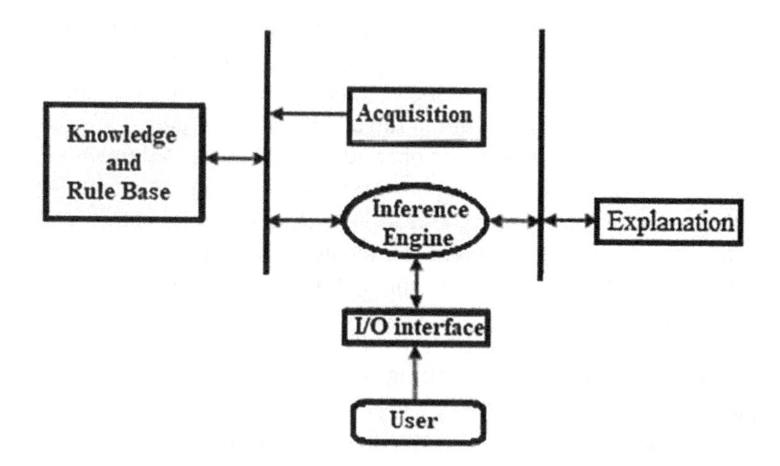

FIGURE 14.3 Expert system.

i. It guides and assists humans while making decisions.
ii. It is easy to understand.
iii. It is reliable and highly responsive.

Disadvantage

Using ES, the following drawbacks or disadvantages occur:

i. ES has a difficulty to adapt or pursue new difficulties or circumstances.

14.2.3.2 Applications of Expert System in Power System

There are many applications of the ES used in the power system, like monitoring system, control system, decision-making etc. ES has computer programs to solve problems having a huge quantity of data and information in a short time duration (Sathiyasekar, Thyagarajah, & Krishnan, 2011).

ES is the type of computer programs. So, we have to write the codes in the simpler than the actual method of calculating the parameters of the power system like distribution, transmission, and generation. Due to the advantage such as easy modification of computer program, this technique is very useful in the power system (Chen, Zhang, & Vachtsevanos, 2012).

14.2.4 Genetic Algorithm (GA)

Genetic algorithm (GA) is the research-based optimization approach based on the analysis of genetics and their selection. GA solves problems to determine the optimal solution of difficult problems, which take a life span to solve. It is based on the principle that the population have the fittest individual having the highest probability and the highest possibility for survival (Cleves, 2008). GA is developed to solve the optimization problem in machine learning and in research.

14.2.4.1 Advantages and Disadvantages of Genetic Algorithm

Using GA for the estimation and prediction purposes have the following advantages and benefits:

i. Genetic algorithm is fast in operation.
ii. Genetic algorithm is efficient
iii. Genetic algorithm optimizes the continuous as well as discrete function
iv. GA is useful when the area of search is very large, and there are huge number of parameters involved.

Disadvantages

Using genetic algorithm, the following drawbacks or disadvantages often occur:

i. Genetic algorithm is not suitable for the simple problem having the derivative information available.
ii. In genetic algorithm, there is no guarantee of the quality and optimality of the solution.

14.2.4.2 Applications of Genetic Algorithm in Power System

If there is any fault on the transmission line, the fault detectors attached along the transmission line detect the fault and forward it to the FLS (Bhargava & Handa, 2018). In the FLS, the input of the fault detector is fed to the fuzzification system, and then the output of this system is fed to the fuzzy system in which the fuzzy rules are implemented on it (Antony, Bardhan Anand, Kumar, & Tiwari, 2006). Then, the output of the fuzzy system is further fed to the defuzzification system, and we get the crisp output or solution to perform the operation after fault. There are some environmental sensors that are attached in the AI system of transmission line, which will sense the atmospheric condition in the output of environment sensors in given to the input of ES and ANN (Zhu & Deshmukh, 2003). The output of ES and ANN is utilizing to increase the performance of the line.

i. **Planning**: Reactive power optimization, wind turbine positioning, network feeds routing, and optimal placement of capacitor.
ii. **Operation**: FACTS devices control, hydro-thermal scheduling, scheduling of maintenance, minimization of losses.
iii. **Analysis**: Load flow analysis, load frequency control (LFC), reduction in harmonics distortion, design of filters.

14.3 COMPARISON OF AI TECHNIQUES IN POWER SYSTEM

The comparison between various AI techniques in the power system is shown in Table 14.1.

TABLE 14.1
Comparison of AI Techniques

	Techniques			
Features	**ANN**	**Fuzzy Logic**	**GA**	**Expert System**
Knowledge used	Training set	Rules database	Selecting, best fitting, optimization	Rules, objects, frames etc.
Self-learning	Natural	Possible	Possible	Possible
Handling vague cases	Natural	Possible	Possible	Possible
Computations	Using hardware	Moderate	Moderate	Extensive
Robustness	Difficult	Easy	Easy	Easy

14.4 APPLICATIONS OF ARTIFICIAL INTELLIGENCE IN POWER SYSTEM

Many problems in the modern power system cannot be evaluated by traditional techniques that are based on the mathematical evaluation. The areas of AI application in the power system are as follows (Hayward & Davidson, 2003):

i. Power system planning like expansion of generation, transmission, and distribution etc.
ii. Power system operations like economic load dispatch, load and power flow analysis, hydro-thermal scheduling, unit commitment, maintenance scheduling.
iii. Power system load frequency like short term, long term, and medium term.
iv. Power system planner at a distributive generation like solar power plant at consumer premises and other renewable energy sources.

14.5 CONCLUSION

Conventionally, planning, controlling, generation, transmission, distribution etc. of the power system were evaluated mathematically. The conventional technique is not accurate and precise but also takes time to calculate, due to a delay in decision-making. While developing the field of power system, the AI techniques are introduced. There are many applications of AI in the power system, which are accurate, efficient, and also takes less time to calculate.AI is helpful in decision-making even in the case of emergency. There is another upcoming technology of AI in the power system, i.e., Smart Grid. Smart Grid is the grid having the ability to control, monitor, plan, and maintain the power system automatically and also the integration of distributive energy generation.

REFERENCES

Antony, J., Bardhan Anand, R., Kumar, M., & Tiwari, M. (2006). Multiple response optimization using Taguchi methodology and neuro-fuzzy based model. *Journal of Manufacturing Technology Management*, 17(7), 908–925.

Bhargava, C., Banga, V. K., & Singh, Y. (2016). Improved reliability, accuracy and failure minimization of spacecraft satellite architecture by using fuzzy logic. *International Journal of Control Theory and Applications*, 9(41), 523–532.

Bhargava, C., & Handa, M. (2018). An intelligent reliability assessment technique for bipolar junction transistor using artificial intelligence techniques. *Pertanika Journal of Science & Technology*, 26(4), 1765–1776.

Chen, C., Zhang, B., & Vachtsevanos, G. (2012). Prediction of machine health condition using neuro-fuzzy and Bayesian algorithms. *IEEE Transactions on Instrumentation and Measurement*, 61(2), 297–306.

Cleves, M. (2008). An introduction to survival analysis using Stata. College Station, TX: Stata Press.

Dekker, R. (1996). Applications of maintenance optimization models: a review and analysis. *Reliability Engineering & System Safety*, 51(3), 229–240.

Diehl, R. (2003). *High-Power Diode Lasers: Fundamentals, Technology, Applications* (Vol. 78). New York, NY: Springer Science & Business Media.

Dylis, D. D., & Priore, M. G. (2001, 22–25 January). A comprehensive reliability assessment tool for electronic systems. Paper presented at the *IEEE Annual Symposium on Reliability and Maintainability*, Philadelphia, PA.

Hayward, G., & Davidson, V. (2003). Fuzzy logic applications. *Analyst*, 128(11), 1304–1306.

Samanta, B., & Al-Balushi, K. (2003). Artificial neural network based fault diagnostics of rolling element bearings using time-domain features. *Mechanical Systems and Signal Processing*, 17(2), 317–328.

Sathiyasekar, K., Thyagarajah, K., & Krishnan, A. (2011). Neuro fuzzy based predict the insulation quality of high voltage rotating machine. *Expert Systems with Applications*, 38(1), 1066–1072.

Stanton, L. G. (2009). Modeling in pattern formation with applications to electrochemical phenomena. (Doctor of Philosphy), Northwestern University.

Zhu, J., & Deshmukh, A. (2003). Application of Bayesian decision networks to life cycle engineering in Green design and manufacturing. *Engineering Applications of Artificial Intelligence*, 16(2), 91–103.

Zhu, W., Xu, L., Pang, J. H., Zhang, X., Poh, E., Sun, Y., . . . Tan, H. (2008). Drop reliability study of PBGA assemblies with SAC305, SAC105 and SAC105-Ni solder ball on Cu-OSP and ENIG surface finish. Paper presented at the *2008 58th Electronic Components and Technology Conference*, Lake Buena Vista, FL.

15 Impact of Artificial Intelligence in the Aviation and Space Sector

Nelvin Chummar Vincent, Roshan Rajesh Bhakar, and Sarath Raj Nadarajan Assari Syamala
Amity University

Jerrin Varghese
The University of New South Wales

CONTENTS

15.1 INTRODUCTION

While the travel industry is being disturbed by the arrival of artificial intelligence (AI), the change is for the better. The aviation industry, when it comes to AI, is often accused of trailing behind when compared to other industries. However, AI is producing revolutionary changes in the company's approach towards data, operations, and revenue flow in the industry. Leading aviation companies around the world are already using AI to improve their efficiency in operations and to increase the customer satisfaction [1]. The introduction of AI in airlines and flight operators has significantly reduced various operational costs by suggesting the optimized usage of fleets and operations [2]. One of the most vital components in air travel is customer service, which begins from the moment a passenger starts thinking of booking a flight and not just when they arrive at the airport. Therefore, airlines are looking at how the implementation of AI can reduce and minimize the impact of any disruptions on the passengers' experience and their business. The airlines now use AI to predict how many items of certain food and drink were required on different flights, both improving the passenger experience and minimizing waste [3]. AI is used by a few airlines for predictive analytics, auto-scheduling, targeted advertising, and customer feedback analysis to enhance the experience of a passenger in their airline [4].

Astrobiologists at NASA hope machine learning can help find data that can be collected by future telescopes and observatories [5]. The AI is especially important as these big datasets from suture observations are going to be very noisy and sparse. Technology along with AI is bringing us closer and closer to the stars and planets. This aspect of data investigation is important for our daily lives and in improving many areas like transport and navigation. Robot with AI aids the process of collection and processing these data so that they can be utilized in necessary areas.

As AI continues to evolve and expand, companies are seeing big changes in their field. Big name brands are investing in AI technology to enhance their products and services to better serve their customers.

15.2 ARTIFICIAL INTELLIGENCE IN AIRLINE PASSENGER IDENTIFICATION

The implementation of AI-based neural networks (NNs) in civil aviation is successfully changing today's air transportation industry. World's leading airliners are employing this technology in their service fields to improve customer's service efficiency and in parallel, to enhance customer's overall experience with the airline. As acknowledging the contemporary civil aviation security concerns around the world, current headway technologies such as AI can pave way for a secure air transport in utility for civilians and military alike. Security check points at airports such as kiosks can be well engineered with AI incorporated with advanced machine learning, mechatronics, human language processing ability, and biometrics (such as fingerprint, pupil, and facial recognition) technologies [6]. Discussing about biometrics, facial recognition can be deployed for identifying customer identity, professional background, and the purpose of stay within the country. The objectives of developing such technology would be ease of passenger flow though kiosk while ensuing the maximum security.

The necessity to develop advanced security check technology can be acknowledged after realizing that the numbers of passengers are expected to be doubled in coming decades. According to the International Air Transport Association (IATA), the numbers of passengers (per head) can be expected to increase twice as that of current number of customers to around 8.2 billion active service users in 2037. Furthermore, the compound annual growth rate (CAGR) is expected to increase 3.5% annually, leading to more crowded/rushed airports [7].

15.2.1 Facial Recognition

Facial recognition technology uses nodal points on the individual's face to verify a customer's identity. This algorithm can be used for security checks and to match passengers to their luggage via smart identification tags such as RFID. Facial recognition algorithm module can be installed on handy drones, which can access restricted areas within the aerodrome of the airport for unauthorized personnel [8]. Moreover, the utility of this module can be put into use to describe and implement another algorithm that uses the motion detection and uniform color to recognize an airport staff. Since 2000, enormous efforts have been made to automate the process of human recognition and tracking. Facial recognition technology paired with AI has shown acceptable results in in situ/controlled environments. However, these results are inconclusive when the technology is to deal with day-to-day situations. A common example in this criterion includes comparing an unrecognized face against an already-identified person's face. Di Sciascio [9] discusses the solution of problem by categorizing the characteristic properties into high-level and low-level characteristic descriptions. High-level characteristics include the properties contained in distinct and unique features contained in objects in the image. Low-level characteristics of the image include visual descriptions of the image like colors, shade, contrast, brightness etc. Di Sciascio argues that the new facial recognition system should be based on recognizing the high-level characteristic descriptions as they are permanent intrinsic description in an image. On the other hand, the low-level characteristic description changes constantly with a change in color contrast, brightness of the image etc.

Many attempts were made after 2000 to develop a world-class facial recognition system. One common facial recognition system is the principal component analysis (PCA). The PCA method compiles special distinctive spatial relationships between the terrain features of the face, for example, the jawline, nose, eye sockets, brow-ridge etc. [10]. However, for developing an improved, advanced, and accurate facial recognition system, convolutional neural networks (CNNs) are used [11]. In this machine learning algorithm, the high-level characteristic description is used by the system. The subject person in the image is recognized completely based on individual body parts and a full descriptive image of the person. The differentiated body parts and the full image of the subject person are linearly chained/stringed together and fed up a multiconnection layer to produce the final representation if the image is in question. The new and enhanced CNN system provides us with 800-pixel/dimension feature representation of the subject image. The CNN is proficient enough to learning the local characteristic high-level descriptions from the input image. An archetypical classifier model of CNN consists of randomly alternating sequences of convolutional and subsampling layers for feature retrieval and extraction in the last layer. The template size of the convolutional layer is in the form of 5×5 matrix, and the template size of the sublayer is of 2×2 matrix. The input image is resolved and accepted in 23×32 matrix resolution. The set F1 consists of six numbers of statistical feature maps, and the third layer set F3 consists of 12 numbers of statistical feature maps. The remaining sets F1 and F4 have the same amount of statistical feature maps as F1 and F3. After being processed by the four superficial layers, the input image is now traversed in the arrow matrix to obtain an output vector of dimensions 240×1.

Finally, this vector is analyzed by the last layer of the module, which is incorporated by the NN algorithm. This layer classifies the input vector into 30 distinct classes. In the convolutional layer set, F4 and the last layer of NNs, the classes are fully connected. Furthermore, the CNN classifier is completely machine-trained using a backpropagation algorithm in the batch mode; this is done to ease machine learning incorporating the learning of C1 and C2 convolutional masks in the superficial layers and learning the connections between different layers in the NN classifier [12]. Figure 15.1 shows the CNN design.

Facial recognition technology is being used to perform the customer identity verification and to match passengers to their luggage through kiosks. Facial recognition has the potential, as with any other formidable tool, to be used both for good and for nefarious purposes. With superior NNs and chips, the system becomes smarter, increasing the available power, which in turn allows an increased number of layers, enabling superior intelligence. This creates a positive feedback loop [13].

15.3 ARTIFICIAL INTELLIGENCE IN AIRLINE BAGGAGE IDENTIFICATION

Baggage security also plays a crucial role in the airport security. In the 1980s and 1990s, expansion of aviation infrastructure such as airports and air-traffic control facilities did not match up with the growing air traffic [14]. This caused an imbalance in the supply and demand chain that ultimately led to overcrowded airports. Overcrowded airports paid little to no attention to personnel luggage security concerns. This was dangerous because it could pave a way for high-level aviation-based criminal activities like smuggling of illegal substances or objects, hijacking and keeping of civilian passengers as hostages, robbery, illegal transportation of illegal explosives without expert supervision carrying unauthorized firearms or sharp objects in the aircraft's cabin, etc. In the aftermath of the September 9, 2001, terrorist attack in New York City, USA, airports became even more crowded amidst

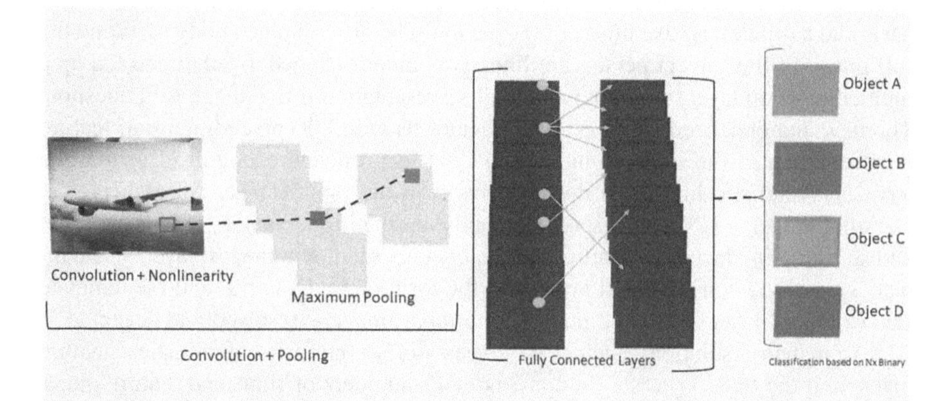

FIGURE 15.1 Convolutional neural network design.

the anxiety oriented towards homeland security [15]. Therefore, the identification of passenger baggage was of utmost importance in the security of civilian passengers and aviation assets. The authorities became ever more aware of the necessity for passenger identification coupled with baggage identification. The United States ramped up its passenger scanning and security efforts to such an extent that passengers were often told to "strip" themselves to make sure that they were safe to travel. But the measures taken by the authorities were far from adequate to prevent another terrorist attack on the United States soil. The reality was that, of more than a billion suitcases, 3.8 million suitcases was scanned in the United States alone on a daily basis. The latest technology available at that time in the country provided the ability to scan around 120–540 bags per hour, or two full-plane loads an hour. To provide the well-assured security, a more reliable, fast, and sophisticated screening technology was needed for the passenger identification and baggage security. Back then, before the 9/11 tragedy, all over the world, normal X-ray machines were used to scan baggage. The responsibility of identifying odd items within baggage rested in the hands of personnel expertise, attentiveness, and the capability and efficiency of the scanning machine. Most X-ray scanning machines had been designed to identify certain hazardous objects just merely based on the virtue of their shape, X-ray permeability, and reflectance. However, these machines consequently failed to identify hazardous materials as indicated by how drug traffickers managed to evade authorities and smuggle products with only marginally more effort than before [16]. Most of the threats easily escaped security protocols merely due to the machine's inability to identify new threats for which it was not programmed.

All security screening technologies in charge of luggage inspection should satisfy a number of criteria including (but not limited to):

i. Satisfactory level of accuracy in object recognition, identification, and categorization.
ii. Quick/fast operation to check airport congestions.

All previous research and literature work on the context of X-ray-based luggage inspection is completely based on the bag of visual words model (BoVW) considering the fact that there is still much more need of imperial evidence in terms of proving that the machines are capable of automatically recognizing and categorizing various objects based on threat they pose [17]. However, the problem has the potential to be solved by incorporating the BoVW with the advance and state-of-the-art algorithm of CNN into the X-ray-based luggage screening and imagery. CNN and BoVW can be connected together in a linear fashion with the help of support vector machine (SVM) classifier [18].

Deep convolutional neural networks (DCNNs) have been widely utilized in many computer chromatography, vision, and imagery applications. Previously as we have seen, we used CNNs in passenger identification utilities at the airport security check post and kiosks. Deep NNs were first proposed by Krizhevesky in Ref. [19]. DCNNs are just like normal CNNs but deeper and wider, have more linear connections, and are more multi-layered than the normal CNNs. Figure 15.2 represents the DCNN architecture.

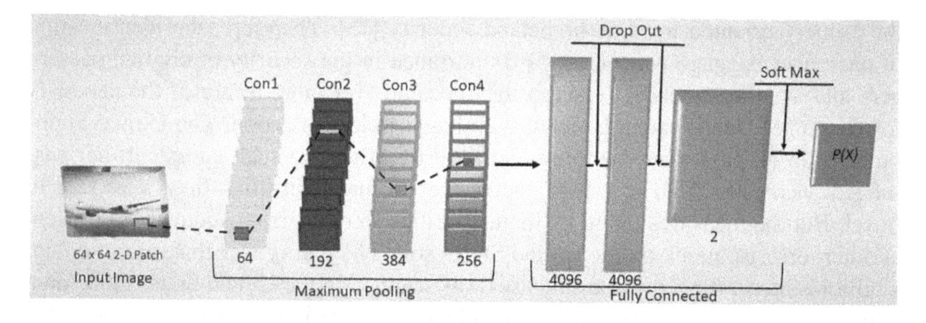

FIGURE 15.2 Deep convolutional neural network architecture.

The DCNNs have around more than four or five primary layers with 11×11 receptive filters and three fc layers in contrast to the normal CNNs that possess just about four layers. These networks provide us with high level of security assurance, as the DCNNs possess high grade of parameterization. However, the use and utility of connection dropouts and hidden neurons are removed in the processing and machining of imagery data to avoid overparameterization and fitting of sparse neural variables such that the performance of the entire machine is dependent on sublayers of the underlying network elements. Moreover, the presence of overparameterized network dropouts and free neurons tend to reduce or degrade the robustness of the present networks to overfitting within the operation linear machine. Likewise, the removal and subsequent reduction of network dropouts and free neurons lead to a nonlinear system, which is definitely far more complex to understand and operate. Thereupon, a novel approach (activation function) was required to enhance DCNNs for the consequent nonlinearity. Consequently, ReLu (rectified linear unit) was introduced in the existing algorithm. In the context of AI and its subdiscipline, ANNs, and the rectified NNs, the rectifier is an "activation function" defined generally by the positive part of its mathematical argument:

$$f(x) = x^{+} = \max(0, x) \tag{15.1}$$

where x is the input to a neuron. This is also known as a ramp function and is analogous to half-wave rectification in electrical engineering. Any unit that makes use of such a rectifier is known as a rectified linear unit (ReLU) [20].

The work regarding ReLU in the DCNNs was majorly accomplished by Zeiler and Fergus [21,22]. Following the work, they designed a similar variant of the model with smaller and shorter receptive fields. Another advancement in the field of DCNNs was done by Simonyan and Zisserman (VGG) in Ref. [23]. In the study, it is shown the significance of the depth of the NNs with convolutional masks stacked up with sort 3×3 receptive fields with step/pace of unity. There are two principal benefits of using small input imagery matrices: first, they increase the nonlinearity of the system by removing the network dropouts, and second, they decrease the overall numbers of parameters in the algorithm in the machine, so this checks the overparameterization of variables and constraints connected to them. Empirical observations have shown that using small 3×3 imagery input matrices, with a carrying depth of 10–20 layers, can pronouncedly yield better performance and efficiency of the system algorithm. Further threat of overparameterization can be reduced by using DCNNs with 50, 101, and 152 filter factorizations that are stacked in smaller input matrices such as 1×1 and 3×3 filters. This was proposed by He et al. in Refs. [24,25].

15.4 ARTIFICIAL INTELLIGENCE IN AIRLINE CUSTOMER SATISFACTION

AI can also become the pathway through which airlines interface with their customers. The user interfaces (UI) of websites now use AI, which attempts to customize the positioning of buttons, font, layouts, etc. of the webpage in response to where the customer seems to be clicking most. These are called intelligent user interfaces (IUI). With this naturally come concerns about whether that means that the machine is influencing the decisions that the user is most likely to take. While this is true, it will generate much stronger user engagement. The grid is a recent example of a product that aimed to customize websites using machine learning [26].

IBM's Watson Assistant and other software can already answer the questions given by a customer (which it interprets through NLU (natural language understanding)) by searching databases for information and can direct particularly complex questions to actual customer representatives [27]. This saves time and money for the organization in the long run. The Dubai Electricity and Water Authority's chatbot, Rammas, answered over 1.2 million customer queries in 2019 [28].

Of the technologies mentioned in this chapter, chatbots are the most ubiquitous. The two common methods of creating a working chatbot are given in the following subsections:

a. Natural Language Processing (NLP)

This consists of NLU and natural language generation (NLG). For NLU, the AI parses the text from the user and converts it into a mix of what it considers meaningful data. It normalizes the user input, correcting common grammatical errors so it can work efficiently. It breaks this text down into "tokens" and checks if these terms can be grouped into word families on the basis of which it can interpret it. These are classified as "intents" (indication of what the user wants) and "entities" (supplementary information). It could identify an address or if the customer is getting emotional and can then hand the chat over to a human representative.

For NLG, the AI first determines what information to include in the text. Then, it structures the arrangement of the text. It aggregates sentences that may have redundant terms for easier reading and makes choices about the lexicon used to describe the concept, e.g., "large" vs "big." Referring expression generation is also important. A referring expression is one that is used to refer to a place, thing, or event, e.g., "3 km west of A" vs "2 km west of B," where B is a lesser-known landmark between A and the destination.

NLG can be created by simply training a machine learning algorithm on large sets of data with the corresponding human-written text [29].

b. Pattern Matching

Here, the program simply searches for patterns in the input and looks for previously fed questions that are similar to it and then sends out appropriate responses. This is easily achieved using Artificial Intelligence Markup Language (AIML), an XML dialect [30]. One could also use AI to monitor what services the user seems more likely to use and increase the prominence of those services in the UI. The SITA Smart Path is a product that has now been used in trials by both Lufthansa and British Airways and that has halved boarding times from 40 minutes to just 20. It automates the boarding process by identifying the face of the passenger after comparing it with Customs and Border Protection data [31].

15.5 ARTIFICIAL INTELLIGENCE IN AIRCRAFT SAFETY AND MAINTENANCE

AI can play a behind the scenes role in customer safety through applications in maintenance and repair operations. Maintenance is one of the top five causes of aircraft delays, so saving time and preventing unplanned maintenance is an investment for the airline. The idea is not particularly novel either.

According to statistics from the US Bureau of Transportation, maintenance has caused about 302 delays in 2019 [32]. This leads to major financial losses for airliners. Having AI that can plan and schedule tasks for maintenance is an investment for the airliner. By the year 2021, about 80% of aerospace and defense industries are expecting to be influenced by the AI-based decision in every level of the department [33].

The most common AI techniques used are case-based reasoning (CBR), genetic algorithm (GA), NN, knowledge-based system (KBS), and fuzzy logic (FL). KBS is one of the oldest approaches taken towards the maintenance management. CBR is a way to add learning abilities to the decision support system (DSS) and for diagnosing faults. GA helps in solving complicated computational processes for optimizing problems. NN is also used for fault diagnosis but also in predictive maintenance. FL deals with uncertainty in the predictive maintenance section.

There are already-working products from Rolls-Royce and Airbus that facilitate the predictive maintenance, i.e., targeting at-risk parts and scheduling servicing based on more accurate estimates of part life. Let us analyze one example to see what AI can achieve in maintenance.

Skywise, from Airbus, first accesses the historical maintenance data of an aircraft or fleet of aircraft. Sensors are installed on all aircraft in sensitive locations, such as the engine, to get the real-time data on how much the life of the part has reduced after each use. Data is uploaded to Skywise and analyzed after the flight. For example, it might notice a slow change in the overall pressure of a hydraulic pump over time and can then flag the part and notify the airline that while it may last for another five flights, failure in another ten is highly likely. If the problem seems particularly bad, the system can also send out a message midflight. This allows the airline to allot more time to the maintenance of the aircraft and gives them time to assign another flight to carry customers in the meantime. The software will also bring up the exact repair manual and procedure using the digitized maintenance manuals [34].

Honeywell's counterpart, the Honeywell Forge (formerly GoDirect Connected Maintenance), reports a no-fault-found (when a part that has been reported faulty is found not to possess any faults) rate of 1.5% and a 35% reduction in operational disruptions [35].

AI is also being used more directly by having them govern maintenance and repair robots. Lufthansa Technik's CAIRE project robot is capable of inspecting fiber-reinforced composites on the skin of an aircraft for damage and then doing the repair itself. The robot scans the damage, identifies the shape of scarf joint and milling path needed, cuts out the damaged part, and then replaces it with the repair layers [36].

Another application of AI is weather analysis to enhance decision-making about flights. Delta Airlines recently launched a proprietary software that uses machine learning to create hypothetical outcomes in anticipation of large-scale disruptions caused by weather. It learns from the impact of weather disruptions to simulate them better over time [37].

The Runway Overrun Prevention System, from Airbus, is another great example. The system calculates whether it is safe to land the plane with the length of runway up ahead at any given moment. This is calculated for both wet and dry runway conditions. It calculates these on a number of static and dynamic parameters, including position, aircraft and engine type, weight, ground speed, outside air temperature, true and calibrated airspeed, wind, center of gravity, and the current slat/flap configuration. It can even tell pilots to use a maximum brake power if it deems the risk of runway overrun too great [38].

Thales is using an AI-enabled system called TopSkySimDebrief for air traffic controller training. The "electronic instructor assistant" monitors everything from the trainee's heart rate to where their eyes are focused during a simulation and even the cadence in their voice. It will alert the instructor if it judges that the person is behaving differently from their baseline [39].

Spark Cognition came up with a solution using AI to predict component failures. It monitors the mechanical systems of the airplane. It can also recommend the corrective actions to be taken against the failures [40].

The major priority of a designer is to improve the fuel efficiency of the airplane. Safety Line, a French company, developed a tool using AI to optimize the flight parameters in order to enhance the consumption of fuel. The fuel consumption is highest when the airplane is in a climb mode. So it improvises the climb profiles of the airplane [41].

Garmin introduced "Telligence," an AI-influenced voice-controlled glass cockpit system. With a tap of a button-and-voice command, the pilot can tune the frequency; change the display screen; and get the traffic data, wind data, etc. Rather than diverting their attention to different tasks, they can focus more on flying [42].

One can easily see that AI in the safety sector of aviation is not just a buzzword people use to describe the future anymore, but technology that is already here.

15.6 ARTIFICIAL INTELLIGENCE INFLUENCE IN REMOTE SENSING

Remote sensing is a useful tool for carrying out geographical analysis. It has made it easier to study and model the depth of knowledge about the Earth and its atmosphere. It covers a tremendous amount of research in spectral, spatial, and temporal analyses. In the past few years, the impact of AI has become a topic of discussion. AI has proved its efficiency over time in many different fields. Elon Musk, the owner of Space X and Tesla, was bold enough to introduce the concept of AI in its cars for public [43]. It is used in the autopilot vision and planning program. The vehicle captures images continuously. It learns, understands, and finds an instant solution in times of difficulties. Similarly, satellite images have a lot of information that needs to be understood and analyzed. It is difficult to obtain information by using mathematical equations for every pixel in the image. It is time-consuming, and the accuracy of the results won't be guaranteed. With the help of machine learning algorithms, data can easily be broken down and sorted according to the requirement [44]. It eases out the entire remote sensing process.

There are ongoing and newer upcoming approaches in different sections of remote sensing. Manifold learning (ML), transfer learning (TL), semi-supervised learning (SSL), active learning (AL) etc., are a few recently introduced machine learning approaches taken towards remote sensing. ML and SL are related together. ML deals with extracting features that are nonlinear and reduce the dimensions; i.e., high-dimensional data is converted to low-dimensional data while keeping the quality factor constant. SSL models the ML data structure. TL updates the land-use dataset based on the temporal result of the dataset on the basis of existing training file, whereas AL utilizes the updated training files [45]. Machine learning has definitely improved areas like object classification, detection, and extraction, coding and onboard processing, etc.

15.6.1 CLASSIFICATION

Classification is subcategorized into two: (a) supervised (SC) and (b) unsupervised (UC). SC uses the concepts of AI like NN and SVM. For classification, a class is created to sort the features. To generate a class for a classified model, it uses the information labeled per pixel (px). For example, a class could be a set of buildings, a set of roads, or a set of vegetation types. Every class is denoted by a color. The classes are differentiated based on the values of each px used to define the class. Every class considers the region of interest (ROI) (i.e., the px points of interest falling under the class). These points help in determining the boundary of the class. Using the concepts mentioned before, SVM helps to vector the boundary of the class. It categories the ROI according to the defined classes. Then, the NN detects the classes based on the provided training input [46–48]. Figure 15.3 shows the classification of a map based on different defined classes [49].

Visualization and monitoring of a similar area of interest (AOI) is a problem faced by UC. There are many ways to perform a UC. A few of them are NN, SVM, and rule-based [50–54]. These three ways are the primary approaches for carrying out the UC. These approaches come under a single term "clustering algorithm." It is a

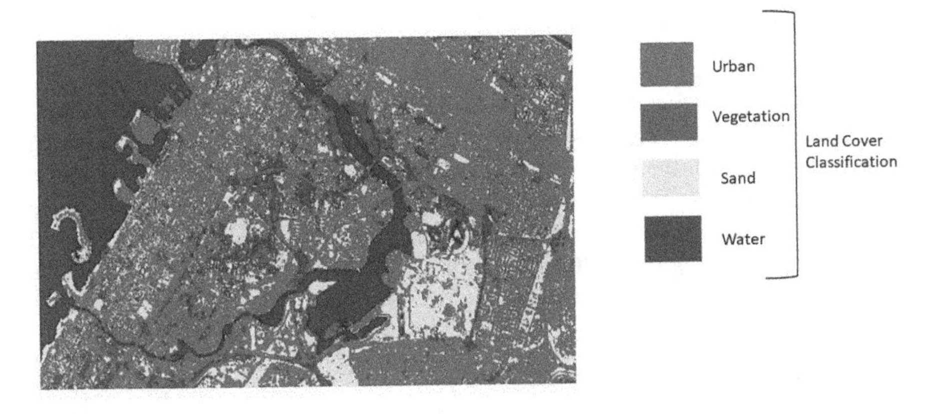

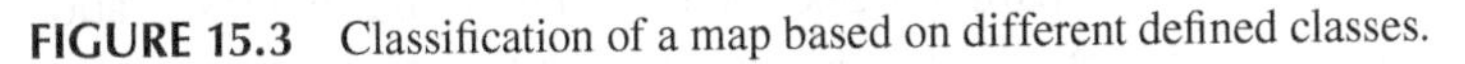

FIGURE 15.3 Classification of a map based on different defined classes.

common statistical analysis method used for UC applications. For generating a UC map, fuzzy clustering is the first step. To exploit the relationship of spatial member, fuzzy clustering is combined with the multiobjective optimization. Contextual regularization (CR) is the next step. It uses methods like fusion of multisource information, projection pursuit, hierarchical clustering, etc. [55–57].

15.6.2 Change Detection

Every satellite orbiting the Earth generates a time series. The time series is generated by an algorithm programmed for the analysis of a particular class. Every time series generated can give an insight for change detection of the class. The larger the time series, the larger would be the required processing time and power. There are direct approaches like change vector analysis (CVA), image subtraction, etc. But these approaches are time-consuming. With the help of the NN and Kernel algorithms, it eases the process of change detection over a long period by considering the SC and UC of the images, and detects the changes for the respective classes [58–60]. Figure 15.4 shows the change detection between three different maps of the same area [61].

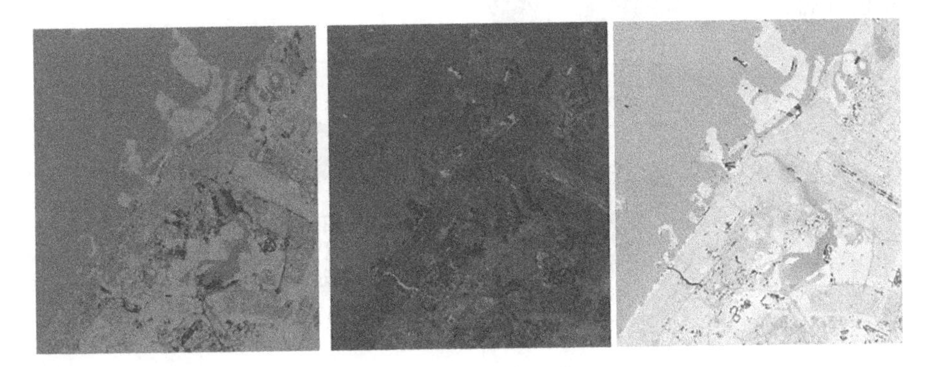

FIGURE 15.4 Change detection between three different maps of the same area.

15.6.3 Feature Extraction

Satellite datasets are large in size, longer processing times are required, and the presence of noise affects the quality of the product. Every band of the dataset has a meaning. Instead of extracting every bit of information from image, machine learning approach can extract the required information only. This reduces the time required to process a dataset. Feature selection uses GAs or SVM approach to eliminate unwanted features. Feature extraction uses linear approaches like PCA and partial least-square (PLS) regression [62–64]. Figure 15.5 shows the extraction of features using wavelengths of absorbed radiation [65].

15.6.4 In-Orbit Image Processing

The ability of autonomous on-board process is trending in Earth Observation Satellite (EOS), where satellites can think independently. UoSat was the first satellite to have an autonomous onboard processor. Currently, European Space Agency and National Aeronautics and Space Administration, the world's largest space agencies, have developed the ability of autonomous onboard processing in their satellites: for example, a small satellite Project for Autonomy On-board (PROBA) developed by ESA and an algorithm for the Autonomous Science-craft experiment software by NASA for the TechSat-21 mission. Instead, Earth Observation-1 satellite used this algorithm as the TechSat 21 mission was aborted [66–70].

The most common functions of an onboard processor are (a) image compression, (b) image analysis, and (c) image selection. Out of the three, image compression is the most important one. Images captured by the EO satellites are very huge in data size. Compression reduces the download time by reducing the size. It uses lossy or

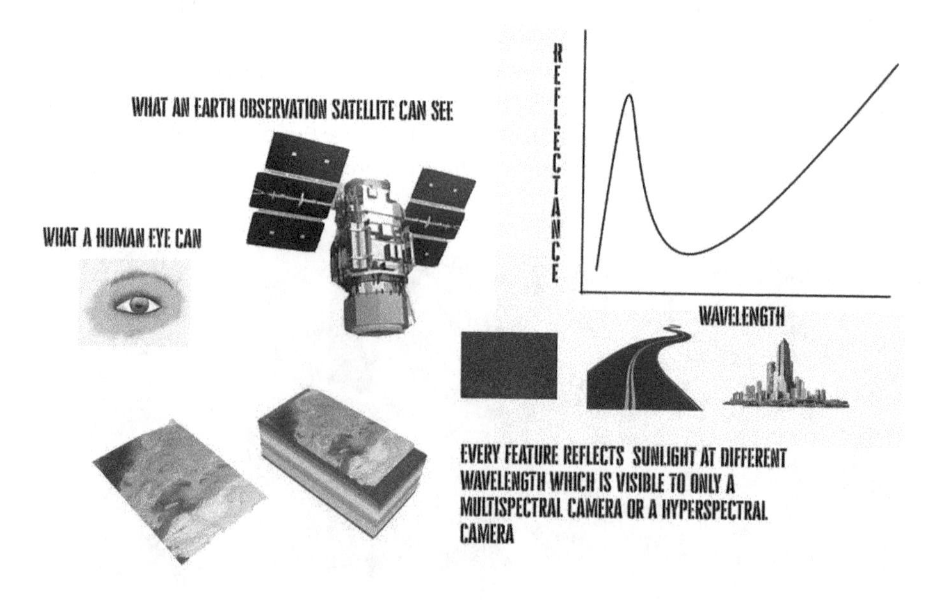

FIGURE 15.5 Extraction of features using wavelengths of absorbed radiation.

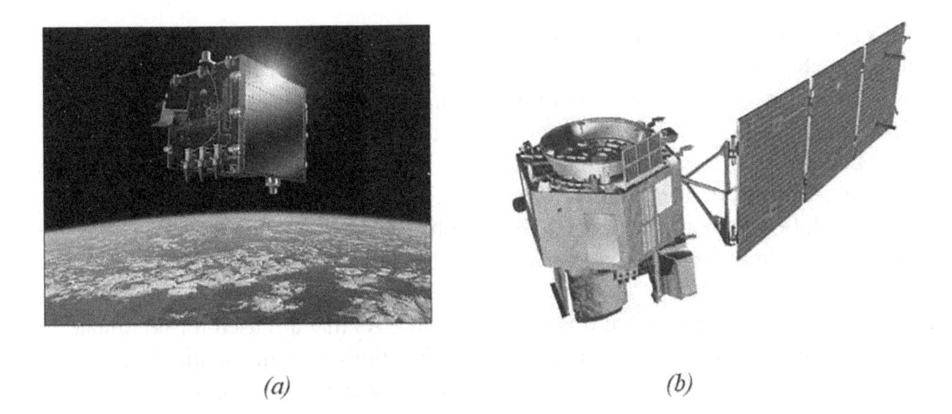

(a) (b)

FIGURE 15.6 (a) PROBA (L) by ESA and (b) Earth Observation-1 (R) by NASA.

lossless. Lossy is a very high ratio compression, whereas lossless compresses the image without losing the quality factor. Autonomous ability can function to analyze, detect, and classify objects according to certain requirements; for example, you have a requirement to detect an amount of vegetation in your area. You can program the onboard processor to preprocess images captured by the satellite. Then, the onboard processor performs the calculation to obtain the statistical data and compare it with the training data to find the percentage of relevance. If the images captured by the sensor don't fit with the requirement, it will be instantly discarded. If they fit well with requirement, the onboard processor can post-process the analysis and prepare the image for downloading [71]. Figure 15.6 shows PROBA (L) by ESA and Earth Observation-1 (R) by NASA [72,73].

15.7 ARTIFICIAL INTELLIGENCE IN SPACECRAFT DYNAMICS

There are many reasons for the introduction of AI in spacecraft. The first is obviously to monitor the condition of the spacecraft and of the environment; it is operating at any given time. This informs further action and whether the crew (if any) and spacecraft are safe. The second would be in planning and scheduling. Optimizing the timing for operations increases efficiency, and the NNs with their ability to spot minute relations between parameters are well suited to this. AI can also be applied during the execution of those plans to ensure that the results are as expected and make use of the spacecraft subsystems as soon as something is off-track. This results in reduced crew workload, and new balances of responsibility between crew, ground staff, and the vehicle [74].

Orientation and orbit control are required for observational satellites and telescopes so they can be pointed at their target at all times. Reorientation maneuvers are usually done by control algorithms based on disturbance torques. When operating satellites around asteroids however, the gravitational disturbance torques are unknown and the gravitational field may be nonlinear. For such cases, one can use the NNs that function to approximate the nonlinear functions and disturbance torques in a real time.

Neural reinforcement algorithms can be used to design orbit controllers that maximize the hovering efficiency around small bodies by modeling the control problem as a Markov decision problem and implementing direct policy search algorithms to find the potential control schemes for the orbit station keeping [75].

AI can also be used to optimize routes for space exploration so as to maximize the fuel efficiency or minimize the time taken. First, we will look at some techniques using evolutionary optimization and their applicability. The following techniques are useful for optimizing for a single objective.

a. **Differential Evolution**: It is a variation in GAs that are used for nonlinear and nondifferentiable continuous space functions, like in chemical propulsion spacecraft transfers, where sequences of multiple impulsive velocity increments must be fixed.
b. **Particle Swarm Optimization**: It is a method where a particle represents a possible solution to the problem. Multiple particles head towards what is currently found to be the best-fitting area found among all the particles (this is updated at the end of each cycle) in the search space for the solution. With each cycle, it gets closer and closer to a true optimal solution, but there is no guarantee that the most optimized solution will be found. It has been used in conjunction with differential evolution algorithms to find the global-optimal solutions consistently for Earth–Mars transfer scenarios.
c. **Covariance Matrix Evolution Strategy**: It modifies the covariance matrix of the problem to ensure a higher probability that the resulting solution is closer to the previously found "useful" solutions.
d. **Ant Colony Optimization**: Like the name suggests, it works similarly to how an ant colony searches for food; "pheromones" are dropped such that a more promising path of possible solutions to explore is given higher priority, and the "distance" to each of those destinations can reduce the priority of any given path [76].

Single-objective optimization is a powerful tool, but one must keep in mind that sometimes, the parameters one is trying to optimize might come into conflict with one another. For example, let us say one wants to build a rocket and minimize the weight, but also maximize the amount of fuel that can be carried. This leads into the topic of **multiobjective optimization.** This requires the concept of Pareto efficiency. In a Pareto-efficient system, the configuration is such that there is no gain one can create in one parameter without creating a loss in another. A Pareto front is thus the collection of all possible configurations to create a Pareto-efficient system. A system is said to be Pareto-dominated if it can still be improved to create a truly Pareto-efficient system.

The **Non-Dominated Sorting Genetic Algorithm II (NGSA-II)** has been used to optimize planetary flyby sequences, minimizing the flight time and fuel consumption. Machine learning, used in conjunction with these evolutionary optimization methods, enables practical, real-time calculations, such as by the application of a NN to find initially optimal conditions in swarm intelligence algorithms for interplanetary transfer, reducing what is usually a high convergence time, using SVMs [77].

It has been shown that deep learning can be trained on the optimal-situation feedback on continuous time, nonlinear, deterministic systems like landing of a spacecraft. It appears that these systems can even learn the underlying principles of the system they are modeling, like the Hamilton–Jacobi–Bellman equations. This can be extended to having the system derive the optimal path for transition from Earth to Mars orbit [78].

Even power systems can now be automated using AI, using ANNs or fuzzy logic controllers. Fuzzification assigns the degrees of membership to input data under certain membership functions. The rules behind assigning membership can depend on several variables. This means a fuzzy logic controller can be applied to both multiple input–multiple output and single input–single output situations. A defuzzifier maps the output fuzzy sets to crisp output values. Let us look at a possible design for a solar cell-based power system. It uses two inputs for the fuzzy logic controller: the first is the error signal indicating the difference between the output generation and the reference load, and the second is the derivative of that error signal. The output of the

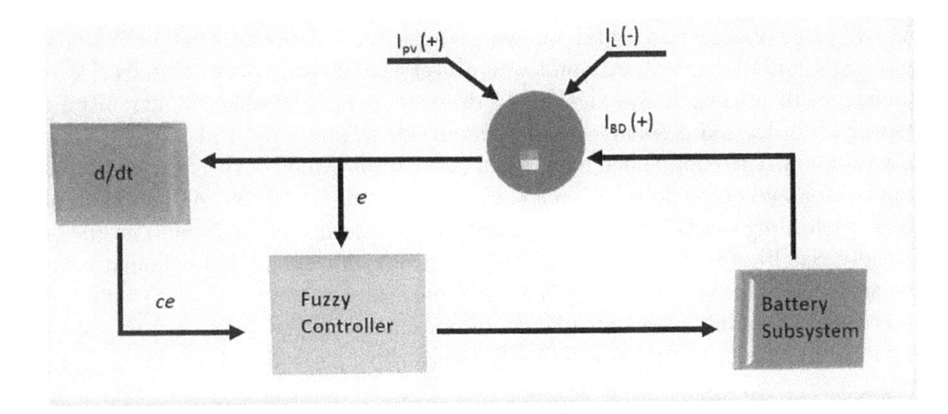

FIGURE 15.7 Block diagram for a fuzzy logic controller.

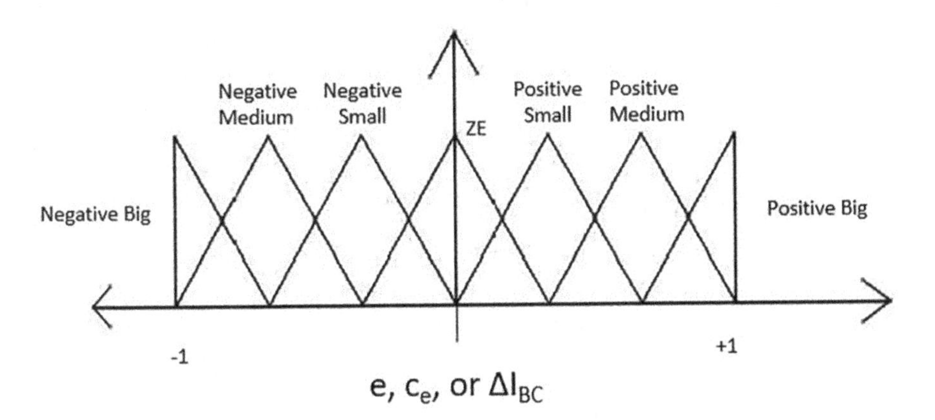

FIGURE 15.8 Triangular membership functions from Negative Big to Positive Big.

controller is the change in battery charge current. The block diagram for a proposed fuzzy logic controller is shown in Figure 15.7.

Standard triangular membership functions are chosen as both the input and the output of the fuzzy logic controller come in the categories: negative big, negative medium, negative small, 0, positive small, positive medium, and positive big. Triangular membership functions from Negative Big to Positive Big are shown in Figure 15.8.

NN controls using backpropagation have been proposed for operating power control units, which control whether the system is in peak power or whether it is in eclipse condition on the basis of the change in battery current. This is desirable because you don't need to feed a self-learning algorithm data about how the system works. Simply let it observe the relations between parameters through simulations and it will learn on its own. The input network parameters here are the load current and error signal, and the output network is the change in battery current [79,80].

15.8 FUTURE PROSPECTS

AI will bring about a new world, one where airport check-ins are completed wordlessly and within minutes. A world where every passenger's query and service is attended with an ease.. In this ideal world, drones now fully capable of distinguishing between hostiles and civilians will work alongside troops, almost like an extension of a battalion. Their automated maintenance and optimization of fuel usage will free up time for soldiers to do their job and concentrate on the present. AI will monitor airspace, leading to a complete clarity about whether an incoming drone is hostile or friendly. Satellites will map out new worlds, identifying terrain and potential landing spots to facilitate human exploration into the solar system. Robots will get rid of busywork and maintain habitats like the ISS or our future planetary colonies, keeping them in a tip-top shape. Rockets will transport people with only optional inputs to get them from one location to another. Analytics about passengers will be at an airline fingertips, including door-to-door travel times and mode of travel, allowing them to optimize the passenger experience. Aircraft designers will be able to conduct generative designs by running through multiple iterations of a product to maximize the efficiency. Aircraft will be able to identify hazards using the machine vision. For us to live in this world would take a number of refinements in our understanding of AI, each one propping up the next. The more people take an interest in the applications of AI, the faster the progress of the aviation industry.

15.9 CONCLUSION

In the aviation Industry, AI is being thoroughly investigated in the commercial sectors of the aviation industry and is widely being implemented in several areas, including customer service, airport, and flight operations. Although the major commercial passenger airlines are fairly early adopters of AI, there is still great potential for innovation and automation to be explored. Till date, we have seen that with AI technology, there is a massive enhancement in the level of air safety and maintenance, self-service solutions, customer experience with automation as well as the optimization of employee workflow. AI has also brought a substantial level of profit and saving as it enables the industry leaders to make informed decisions by smartly using data prior to making decisions about market positioning and pricing.

AI has presented several opportunities in the field of aerospace, how it is still in its very early phase, especially considering the stringent level of safety measures that need to be maintained for such a strictly regulated industry. Each technology or system that is developed needs to be validated and certified extensively, and often may not be approved by international FAA processes, and therefore, superior verification methods are required in order for AI to attain its maximum potential. Another major barrier that the space industry needs to overcome is the data privacy concerns and cybersecurity practices that have to be duly considered while developing any systems for AI.

As applications for AI in the space and aviation industry continue to broaden, more competitors and leaders are eager to implement these AI-powered solutions. Although there is a substantial level of investment and effort required to the wider adoption of AI, it is accepted worldwide that this innovative technology has a significant potential that can largely optimize manufacturing, provide increased performance as well as efficiently deal with malfunctions.

REFERENCES

1. Leong, L.Y., Hew, T.S., Lee, V.H. and Ooi, K.B., 2015. An SEM–artificial-neural-network analysis of the relationships between SERVPERF, customer satisfaction and loyalty among low-cost and full-service airline. *Expert Systems with Applications*, 42(19), pp. 6620–6634.
2. Choi, S., Kim, Y.J., Briceno, S. and Mavris, D., 2017, September. Cost-sensitive prediction of airline delays using machine learning. In *2017 IEEE/AIAA 36th Digital Avionics Systems Conference (DASC)* (pp. 1–8). IEEE, St. Petersburg, FL.
3. Aviation Business News, 2020. How Artificial Intelligence is Supporting the Aviation Industry. [Online] Available at: https://www.aviationbusinessnews.com/low-cost/artificial-intelligence-aviation-industry/ [Accessed 1 May 2020].
4. Barros, C.P. and Wanke, P., 2015. An analysis of African airlines efficiency with two-stage TOPSIS and neural networks. *Journal of Air Transport Management*, 44, pp. 90–102.
5. NASA, 2020. NASA Applying AI Technologies to Problems in Space Science. [Online] Available at: https://www.nasa.gov/feature/goddard/2019/nasa-takes-a-cue-from-silicon-valley-to-hatch-artificial-intelligence-technologies [Accessed 1 May 2020].
6. Yanushkevich, S.N., 2006, July. Synthetic biometrics: a survey. In *The 2006 IEEE International Joint Conference on Neural Network Proceedings* (pp. 676–683). IEEE, Vancouver, BC, Canada.
7. Iata.org, 2020. COMUNICADO No: 62. [Online] Available at: https://www.iata.org/contentassets/db9e20ee48174906aba13acb6ed35e19/2018-10-24-02-sp.pdf [Accessed 29 April 2020].
8. Walker, S., Bruce, V. and O'Malley, C., 1995. Facial identity and facial speech processing: Familiar faces and voices in the McGurk effect. *Perception & Psychophysics*, 57(8), pp. 1124–1133.
9. Celentano, A. and Di Sciascio, E., 1998. Feature integration and relevance feedback analysis in image similarity evaluation. *Journal of Electronic Imaging*, 7(2), pp. 308–317.
10. Agrawal, B., Dubey, S. and Dixit, M., 2020. Standard statistical feature analysis of image features for facial images using principal component analysis and its comparative study with independent component analysis. In *International Conference on Intelligent Computing and Smart Communication 2019* (pp. 1027–1048). Springer, Singapore.

11. Kokoulin, A.N., Tur, A.I., Yuzhakov, A.A. and Knyazev, A.I., 2019, January. Hierarchical convolutional neural network architecture in distributed facial recognition system. In *2019 IEEE Conference of Russian Young Researchers in Electrical and Electronic Engineering (EIConRus)* (pp. 258–262). IEEE, Saint Petersburg and Moscow, Russia.
12. Kim, J., Kim, J., Jang, G.J. and Lee, M., 2017. Fast learning method for convolutional neural networks using extreme learning machine and its application to lane detection. *Neural Networks*, 87, pp. 109–121.
13. Baker, J. and Baker, J., 2020. AI At Airports: How is Artificial Intelligence Speeding Up Security?. *Airport Technology*. [Online] Available at: https://www.airport-technology.com/features/ai-at-airports-security/?cv=1 [Accessed 1 May 2020].
14. Reynolds-Feighan, A.J. and Button, K.J., 1999. An assessment of the capacity and congestion levels at European airports. *Journal of Air Transport Management*, 5(3), pp. 113–134.
15. Muthukkumarasamy, V., Blumenstein, M., Jo, J., Green, S., 2004. Intelligent illicit object detection system for enhanced aviation security. In *International Conference on Simulated Evolution and Learning*.
16. Reginelli, A., Russo, A., Urraro, F., Maresca, D., Martiniello, C., D'Andrea, A., Brunese, L. and Pinto, A., 2015. Imaging of body packing: errors and medico-legal issues. *Abdominal Imaging*, 40(7), pp. 2127–2142.
17. Kundegorski, M.E., Akçay, S., Devereux, M., Mouton, A. and Breckon, T.P., 2016. On using feature descriptors as visual words for object detection within x-ray baggage security screening. In *International Conference on Imaging for Crime Detection and Prevention, IET* (pp. 1–12).
18. Bouamra, W., Djeddi, C., Nini, B., Diaz, M. and Siddiqi, I., 2018. Towards the design of an offline signature verifier based on a small number of genuine samples for training. *Expert Systems with Applications*, 107, pp. 182–195.
19. Krizhevsky, A., Sutskever, I. and Hinton, G.E., 2012. Imagenet classification with deep convolutional neural networks. Advances in *Neural Information Processing Systems*, 25, pp. 1097–1105. https://kr.nvidia.com/content/tesla/pdf/machine-learning/imagenet-classification-with-deep-convolutional-nn.pdf
20. Zhang, Z., Brand, M. and Mitsubishi Electric Research Laboratories Inc., 2019. Machine Learning via Double Layer Optimization. U.S. Patent Application 15/814,568.
21. Hoffer, E. and Ailon, N., 2015, October. Deep metric learning using triplet network. In Feragen, A., Pelillo, M. and Loog, M. (eds.) *International Workshop on Similarity-Based Pattern Recognition* (pp. 84–92). Springer, Cham.
22. Zeiler, M.D. and Fergus, R., 2013. Visualizing and understanding convolutional networks. CoRR, abs/1311.2901. arXiv preprint arXiv:1311.2901. https://link.springer.com/content/pdf/10.1007/978-3-319-10590-1_53.pdf
23. Simonyan, K. and Zisserman, A., 2014. Very Deep Convolutional Networks for Large-Scale Image Recognition. CoRR, vol. abs/1409.1556. [Online] Available at: http://arxiv.org/abs/1409.1556.
24. He, K., Zhang, X., Ren, S. and Sun, J., 2016. Deep residual learning for image recognition. In *Proceedings of the IEEE Conference on Computer Vision and Pattern Recognition* (pp. 770–778). IEEE, Las Vegas, NV.
25. Szegedy, C., Vanhoucke, V., Ioffe, S., Shlens, J. and Wojna, Z., 2015. Rethinking the Inception Architecture for Computer Vision. CoRR, vol. abs/1512.00567. [Online] Available at: http://arxiv.org/abs/1512.00567.
26. Thegrid.io, 2020. The Grid. [Online] Available at: https://thegrid.io/ [Accessed 12 May 2020].
27. High, R., 2012. *The Era of Cognitive Systems: An Inside Look at IBM Watson and How It Works*. IBM Corporation, Redbooks, Armonk, NY, pp. 1–16.

28. Dewa.gov.ae, 2020. Dubai Electricity & Water Authority (DEWA), DEWA'S Virtual Employee Rammas Responds to over 1.2 Million Queries in 2019. [Online] Available at: https://www.dewa.gov.ae/en/about-us/media-publications/latest-news/2020/03/dewas-virtual-employee-rammas [Accessed 12 May 2020].
29. Keselj, V., 2009. *Speech and Language Processing Daniel Jurafsky and James H. Martin (Stanford University and University of Colorado at Boulder).* Pearson Prentice Hall, xxxi+988 pp.; hardbound, ISBN 978-0-13-187321-6.
30. Satu, M.S. and Parvez, M.H., 2015. Review of integrated applications with AIML based chatbot. In *2015 International Conference on Computer and Information Engineering (ICCIE)*, Rajshahi, pp. 87–90, doi: 10.1109/CCIE.2015.7399324.
31. Passenger Self Service, 2020. Lufthansa Now Using Biometric Boarding at Miami. *Passenger Self Service.* [Online] Available at: http://www.passengerselfservice.com/2019/02/lufthansa-now-using-biometric-boarding-at-miami/ [Accessed 12 May 2020].
32. Tribune, A., 2020. How Airlines are Using Artificial Intelligence to Optimize Maintenance Operations. *Aviation Tribune, Aviation News.* [Online] Available at: https://aviationtribune.com/featured-content/how-airlines-are-using-artificial-intelligence-to-optimize-maintenance-operations/ [Accessed 28 April 2020].
33. AMFG, 2020. AI & Aerospace: 5 Ways Artificial Intelligence Could Impact Aviation. AMFG. [Online] Available at: https://amfg.ai/2018/08/31/ai-aerospace-5-ways-artificial-intelligence-could-impact-aviation/ [Accessed 11 April 2020].
34. Iata.org, 2020. Opening Remarks. [Online] Available at: https://www.iata.org/contentassets/c1258b62226b43a0a0f46ce27daf820b/ads-2019-26jun-safety-flightoperations.pdf [Accessed 12 May 2020].
35. Aerospace.honeywell.com, 2020. Godirect™ Connected Maintenance. [Online] Available at: https://aerospace.honeywell.com/content/dam/aero/en-us/documents/learn/services/maintenance-and-service-plans/brochures/N61-1905-000-001_ConnectedMaintenance-br.pdf [Accessed 13 May 2020].
36. Breuer, U., 2016. *Repair. Commercial Aircraft Composite Technology.* Springer, New York, NY, pp. 141–162.
37. Avionics, 2020. Delta Develops Artificial Intelligence Tool To Address Weather Disruption, Improve Flight Operations. *Avionics.* [Online] Available at: https://www.aviationtoday.com/2020/01/08/delta-develops-ai-tool-address-weather-disruption-improve-flight-operations/ [Accessed 13 May 2020].
38. Villaumé, F. and Lagaillarde, T., 2015. Fast: runway overrun prevention system (ROPS). Airbus, Technical Report, p. 55.
39. Avionics, 2020. Thales is Using AI to Augment Air Traffic Management. *Avionics.* [Online] Available at: https://www.aviationtoday.com/2019/01/24/thales-using-ai-augment-atm/ [Accessed 13 May 2020].
40. SparkCognition, Inc., 2020. AI in Aviation Industry, Artificial Intelligence in Aviation. [Online] Available at: https://www.sparkcognition.com/industries/aviation/ [Accessed 29 April 2020].
41. Safety-Line, 2020. Safety Line Pitches Optimised Climb Profile To Save Fuel (Flight Global). *Safety-Line.* [Online] Available at: https://www.safety-line.fr/en/safety-line-pitches-optimised-climb-profile-to-save-fuel-flight-global/ [Accessed 2 May 2020].
42. Garmin Blog, 2020. Talking to Your Airplane with Telligence Voice Control. *Garmin Blog.* [Online] Available at: https://www.garmin.com/en-US/blog/aviation/talking-to-your-airplane/ [Accessed 18 April 2020].
43. Tesla, 2020. Tesla Autopilot. [Online] Available at: https://www.tesla.com/en_AE/autopilot [Accessed 3 May 2020].
44. Domingos, P., 2012. A few useful things to know about machine learning. *Communications of the ACM*, 55(10), pp. 78–87.

45. Camps-Valls, G., 2009, September. Machine learning in remote sensing data processing. In *2009 IEEE International Workshop on Machine Learning for Signal Processing* (pp. 1–6). IEEE, Grenoble, France.
46. Benediktsson, J.A., Palmason, J.A. and Sveinsson, J.R., 2005. Classification of hyperspectral data from urban areas based on extended morphological profiles. *IEEE Transactions on Geoscience and Remote Sensing*, 43(3), pp. 480–491.
47. Del Frate, F., Pacifici, F., Schiavon, G. and Solimini, C., 2007. Use of neural networks for automatic classification from high-resolution images. *IEEE Transactions on Geoscience and Remote Sensing*, 45(4), pp. 800–809.
48. Camps-Valls, G. and Bruzzone, L., 2005. Kernel-based methods for hyperspectral image classification. *IEEE Transactions on Geoscience and Remote Sensing*, 43(6), pp. 1351–1362.
49. Wang, C., Xu, A., Li, X., 2018. Supervised classification high-resolution remote-sensing image based on interval type-2 fuzzy membership function. *Remote Sensing*, 10, 710.
50. Baraldi, A., Puzzolo, V., Blonda, P., Bruzzone, L. and Tarantino, C., 2006. Automatic spectral rule-based preliminary mapping of calibrated Landsat TM and ETM+ images. *IEEE Transactions on Geoscience and Remote Sensing*, 44(9), pp. 2563–2586.
51. Zhong, Y., Zhang, L., Huang, B. and Li, P., 2006. An unsupervised artificial immune classifier for multi/hyperspectral remote sensing imagery. *IEEE Transactions on Geoscience and Remote Sensing*, 44(2), pp. 420–431.
52. Baraldi, A. and Parmiggiani, F., 1995. A neural network for unsupervised categorization of multivalued input patterns: an application to satellite image clustering. *IEEE Transactions on Geoscience and Remote Sensing*, 33(2), pp. 305–316.
53. Awad, M., Chehdi, K. and Nasri, A., 2007. Multicomponent image segmentation using a genetic algorithm and artificial neural network. *IEEE Geoscience and Remote Sensing Letters*, 4(4), pp. 571–575.
54. Mukhopadhyay, A. and Maulik, U., 2009. Unsupervised pixel classification in satellite imagery using multiobjective fuzzy clustering combined with SVM classifier. *IEEE Transactions on Geoscience and Remote Sensing*, 47(4), pp. 1132–1138.
55. Bachmann, C.M., Donato, T.F., Lamela, G.M., Rhea, W.J., Bettenhausen, M.H., Fusina, R.A., Du Bois, K.R., Porter, J.H. and Truitt, B.R., 2002. Automatic classification of land cover on Smith Island, VA, using HyMAP imagery. *IEEE Transactions on Geoscience and Remote Sensing*, 40(10), pp. 2313–2330.
56. Sarkar, A., Biswas, M.K., Kartikeyan, B., Kumar, V., Majumder, K.L. and Pal, D.K., 2002. A MRF model-based segmentation approach to classification for multispectral imagery. *IEEE Transactions on Geoscience and Remote Sensing*, 40(5), pp.1102–1113.
57. Aleksandrowicz, S., Turlej, K., Lewiński, S., Bochenek, Z., 2014. Change Detection algorithm for the production of land cover change maps over the European Union countries. *Remote Sensing*, 6, pp. 5976–5994, doi: 10.3390/rs6075976.
58. Li, J. and Narayanan, R.M., 2003. A shape-based approach to change detection of lakes using time series remote sensing images. *IEEE Transactions on Geoscience and Remote Sensing*, 41(11), pp. 2466–2477.
59. Kushardono, D., Eukue, K., Shimoda, H. and Sakata, T., 1995, July. Comparison of multi-temporal image classification methods. In *1995 International Geoscience and Remote Sensing Symposium, IGARSS'95. Quantitative Remote Sensing for Science and Applications* (Vol. 2, pp. 1282–1284). IEEE, Firenze, Italy.
60. Lambin, E.F. and Strahlers, A.H., 1994. Change-vector analysis in multitemporal space: a tool to detect and categorize land-cover change processes using high temporal-resolution satellite data. *Remote Sensing of Environment*, 48(2), pp. 231–244.
61. Marçal, A.R. and Castro, L., 2005. Hierarchical clustering of multispectral images using combined spectral and spatial criteria. *IEEE Geoscience and Remote Sensing Letters*, 2(1), pp. 59–63.

62. Pal, M., 2006. Support vector machine-based feature selection for land cover classification: a case study with DAIS hyperspectral data. *International Journal of Remote Sensing*, 27(14), pp. 2877–2894.
63. Archibald, R. and Fann, G., 2007. Feature selection and classification of hyperspectral images with support vector machines. *IEEE Geoscience and Remote Sensing Letters*, 4(4), pp. 674–677.
64. Shahabi, H., Ahmad, B.B., Mokhtari, M.H. and Zadeh, M.A., 2012. Detection of urban irregular development and green space destruction using normalized difference vegetation index (NDVI), principal component analysis (PCA) and post classification methods: A case study of Saqqez city. *International Journal of Physical Sciences*, 7(17), pp. 2587–2595.
65. Liao, W., 2012. Feature extraction and classification for hyperspectral remote sensing images (Doctoral dissertation, Ghent University).
66. Bretschneider, T., 2003. Singapore's Satellite Mission X-Sat. In *4th IAA Symposium on Small Satellites for Earth Observation*, Berlin, Germany.
67. Othman, M., and Arshad, A., 2001. *TiungSAT-1: From Inception to Inauguration.* Astronautic Technology (M) Sdn Bhd, Selangor.
68. Haller, W., BrieB, M., Schlicker, M., Skrbek, W. and Venus, H., 2002. Autonomous onboard classification experiment for the satellite BIRD. In *Proceedings of ISPRS Commission 1 FIEOS Conference*, November 10–15, 2002 (Vol. XXXIV, Part 1), Denver, CO.
69. Yuhaniz, S.S., 2008. Intelligent image processing on board small earth observation satellites. Doctoral dissertation, University of Surrey, United Kingdom. https://epubs.surrey.ac.uk/843084/7/Yuhaniz_10130949.pdf
70. Bernaerts, D., Teston, F. and Bermyn, J., 2004. Proba. *European Space Agency.* [Online]. Available at: http://esamultimedia.esa.int/docs/paper cnes la baule June 2000.pdf.
71. Harrison, J.I. and Tiggeler, A.B., 1999. Evolution of on-board data handling on small satellites in Surrey. In *Proceedings of Data Systems in Aerospace, ESA Eurospace SP447* (pp. 409–411), Lisbon, Portugal.
72. Proba-V Mission and Sensor Description, 2020. Proba-V Mission and Sensor Description. [Online] Available at: https://earth.esa.int/web/sppa/mission-performance/esa-3rd-party-missions/proba-v.
73. NASA, 2020. Earth Observing Mission 1 (EO-1). *NASA.* [Online] Available at: https://www.nasa.gov/directorates/heo/scan/services/missions/earth/EO1.html.
74. Frank, J., 2020. Enabling Autonomous Space Mission Operations with Artificial Intelligence. *Ntrs.nasa.gov.* [Online] Available at: https://ntrs.nasa.gov/archive/nasa/casi.ntrs.nasa.gov/20170011131.pdf [Accessed 2 May 2020].
75. Kumar, K.D. and Muthusamy, V., 2017. Fault tolerant control, artificial intelligence and predictive analytics for aerospace systems: an overview. *Information, Communication and Computing Technology*, 351–362, doi: 10.1007/978-981-10-6544-6_32.
76. Izzo, D., Märtens, M., Pan, B., 2019. A survey on artificial intelligence trends in spacecraft guidance dynamics and control. *Astrodyn*, 3, 287–299.
77. Cassioli, A., Di Lorenzo, D., Locatelli, M., Schoen, F., Sciandrone, M., 2012. Machine learning for global optimization. *Computational Optimization and Applications*, 51(1): 279–303.
78. Izzo, D., Sprague, C., Tailor, D., 2018. Machine learning and evolutionary techniques in interplanetary trajectory design. arXiv preprint arXiv:1802.00180.
79. El-Madany, H.T., Fahmy, F.H., El-Rahman, N.M. and Dorrah, H.T., 2011. Artificial intelligence techniques for controlling spacecraft power system. *World Academy of Science, Engineering and Technology*, 49(1), pp. 546–551.
80. Ismail, M., Dlyma, R., Elrakaybi, A., Ahmed, R. and Habibi, S., 2017, June. Battery state of charge estimation using an artificial neural network. In *2017 IEEE Transportation Electrification Conference and Expo (ITEC)* (pp. 342–349). IEEE, Chicago, IL.

16 Artificial Intelligence for Weather Forecasting

Laxmi A. Bewoor
V.I.I.T., Vishwakarma Institute of Information Technology

Anand Bewoor
Cummins College of Engineering for Women

Ravinder Kumar
Lovely Professional University

CONTENTS

16.1 INTRODUCTION

Weather affects enormously our daily life. There are many applications that require an accurate weather prediction. Some of the applications are agriculture and production, military operations, aviation industry, transportation, communication, and so forth. Weather forecasting is a canonical predictive challenge, which requires expertise in multiple disciplines [1,2]. Dynamic and chaotic nature of atmosphere makes prediction more complex task. It requires more computational power to solve the complex mathematical equations that are related to atmosphere. This opens new research era in weather forecasting, which will perform better in terms of computational cost and accuracy of prediction. Because of such a significant application, researchers are inclined towards this field by predicting weather parameters like humidity, pressure, wind speed, wind direction, evaporation, pollution etc. [3–8]. Among all these parameters, temperature is very essential parameter as many other parameters rely on it. Additionally, it has a major impact on not only human life but also region-wise agricultural activities and utility companies for estimating the demand over upcoming days etc.

Nowadays almost all the weather stations are connected to the internet, which provides a real-time meteorological data [9]. Real-time data is something which is delivered instantly after its collection. Real-time systems are being used worldwide for applications such as traffic monitoring, news content delivery, fraud detection in financial domain, weather prediction, etc. So, eventually predicting the real-time temperature is a crucial challenge.

Weather forecasting methods can be broadly categorized into empirical models and predictive models. Empirical models are computational-intensive, whereas predictive models are data-intensive. Because of the growing use of computers, internet data is generated enormously, and due to easy availability of chip processing power and affordable data storage techniques, predictive models of forecasting are quite popular among researchers as compared with empirical models as they depend totally on data and develop the realistic model.

Various techniques like linear regression, autoregression, multilayer perceptron (MLP), and radial basis function (RBF) networks are applied to predict the atmospheric parameters like temperature, wind speed, rainfall, meteorological pollution etc. [10]. Along with this, data mining tools like support vector machine (SVM), decision trees (CART), artificial neural network (ANN), and clusters are being utilized to explore hidden patterns and trends in such a huge and static meteorological dataset. However, data mining methods have some limitations when they are used for a quick analysis of data streams, which are nonstationary, chaotic, and inconsistent, or have a concept drift. Weather comes under this category as its nature is nonstationary. The aforementioned techniques are quite suitable for static data but the real-time data predicted with such technique is a real motivation and a big challenge.

The subsequent sections will provide the details about the related work in the field of weather prediction using machine learning models along with the limitations and deep learning model used for predicting the real-time temperature.

16.2 RELATED WORK

In order to predict the trend in weather, time is the important factor, and hence, time series is an essential concept in this context. It is essential to know the three major characteristics of time series, viz., autocorrelation, seasonality, and stationarity. Stationary time series is ideal for modeling but getting such a model is really very difficult so different transformations like moving average (MA), exponential smoothing (ES), and autoregressive integrated moving average (ARIMA) are used for building a better predictive model. This section describes various machine learning models that predict the weather data.

16.2.1 Multiple Linear Regression Model (MLR)

MLR describes the linear relationship between set of independent variables (X_1, X_2, … X_k) and independent variable (Y), and is generally represented as

$$Y = \beta_0 + \beta_1 X_1 + \beta_2 X_2 + \ldots + \beta_k X_k + \varepsilon \tag{16.1}$$

Here, Y is predictand and X_1, X_2, … X_k are predictors; β_0, β_1, … β_k are called as the regression coefficient, and ε is the standard error term. In order to develop the model, regression coefficients are calculated, and then, the value of Y is calculated. Actual and predicted results are compared to identify the error term.

This simple model is rigorously used in many forecasting processes. Paras and Mathur [11] developed MLR-based statistical model and predicted various weather parameters. For forecasting, the weather conditions for a particular station were collected locally. The data was processed to obtain some statistical indicators to extract the hidden information present in the time series. These statistical indicators, viz., MA, exponential moving average (EMA), rate of change (ROC), OSCillator (OSC), moments (μ_2, μ_3 and μ_4), and coefficients of skewness and kurtosis, were calculated over certain periods. On the basis of correlation, the features were chosen as inputs to the models and regression equations were obtained for predicting the weather parameters. Sreehari et al. [12] developed a MLR-based model for the rainfall prediction. Amral et al. [13] developed the MLR-based model to predict the demand of power system for the short-term (up to 24-hour) load for the south Sulawesi region. Looking at the ease of this model, this model is easy to implement; however, this model is quite suitable if there exists a linear relationship between predictors and predictand; in the real-time environment, all variables have a nonlinear relationship. Additionally, missing data or incorrect values may hamper the accuracy of the model, and hence, ANN-based models are gaining more attention.

16.2.2 Artificial Neural Network (ANN)

The basic computational unit in a neural network is the neuron or perceptron simulating the behavior of human brain. The neuron inputs correspond to the values of weather predictor variables. The hidden-layer and the output-layer neurons accept an arbitrary number of inputs depending on the synapse type chosen to interconnect the neurons. The most common strategy is to employ weighted synapses, which connect every neuron in the source layer to every neuron in the target layer to form a feed-forward network. The activation function component scales the output from a hidden layer to a useful value. The computed value is compared against the actual value, and the error term is calculated and backpropagated to the entire network, and this type of algorithm is called the backpropagation algorithm.

Kumar et al. [14] reviewed the work done by various researchers using ANN, developed the ANN model, which reduces postprocessing data cost by increasing hidden layers, and observed the behavior of increased hidden layers on the performance and generalization of model. Chattopadhyay and Chattopadhyay [15] proposed an ANN model for monsoon season precipitation prediction for Indian meteorological data. For carrying out this experiment, only rainfall time-series data has been used. ANN-based model proved to be better than MLR model but still it was not suitable for the real-time weather forecasting. Hence, deep learning-based models are gaining much attention nowadays. Shabana and Bewoor [16] reviewed in detail various deep learning models used for the real-time data analysis. The subsequent section reveals about these deep learning models.

16.2.3 Deep Learning Models

Feed-forward neural networks produce very good results when they are applied to solve classification problems, but can fail when it comes to the time-series data. The reason behind this is that they incorporate a static mapping of the input vector to the output vector. However, temporal problems consist of dynamically changing states to be learned during the model representation. One of the examples of this underlying temporal nature is weather forecast, where it is needed to consider the past weather parameters into account while predicting the future weather conditions. These problems are also called as sequence prediction problems, where the input to the neural network is a sequence of feature vectors and the expected output is the correct prediction of the upcoming sequence parameter.

In sequence prediction problems, the presentation of the neural network input vector of specific sequence happens during distinct time steps. This is the root cause behind sequence prediction problems that entail the substitution of static feed-forward neural network with the neural network model that reveals a nonlinear dynamic behavior. Recurrent neural network (RNN) possesses this property.

16.2.3.1 Recurrent Neural Networks

RNN reveals highly nonlinear dynamic behavior as it is the most powerful and robust type of the neural network architecture. RNN is a natural generalization of the feed-forward neural network to sequences. RNN architecture contains a loop in the network, which helps in information persistence. RNNs are the only networks that have an internal memory. The reason behind preferring RNN for the sequential data prediction is RNNs that are capable of remembering important points about the input provided to them. This property enables RNN to be very specific in the next value prediction. RNN is successfully being implemented for sequential data, viz., time series, voice, text, video, audio, speech, financial information, news content, weather, and so forth. RNN can construct a considerable intense understanding of sequence and its context. Therefore, wherever timestamp of each data frame in a sequence of data is more important than the structural content of individual frame, RNN needs to be applied.

Sequential data is just continuous data, where interconnected or same type of things follows each other. Some of the examples of such data are DNA sequence or financial information. There is a very popular type of sequential data, i.e., time-series data, which is just a series of data points ordered in time. Traditional neural networks like feed-forward neural network were not capable of information persistence. RNN manages this issue with the help of loops in their network.

In Figure 16.1, a neural network takes some input x_t and gives the output value h_t. There is a loop in the network at point A, which is the reason for information passing from one step to another step.

From Figure 16.2, it is clear that RNN is nothing but multiple repetitions of the same neural network, where every network carries forward the message to the descendant.

Current state of network is calculated with the following formula:

$$h_t = f\left(h_{t-1}, x_t\right) \tag{16.2}$$

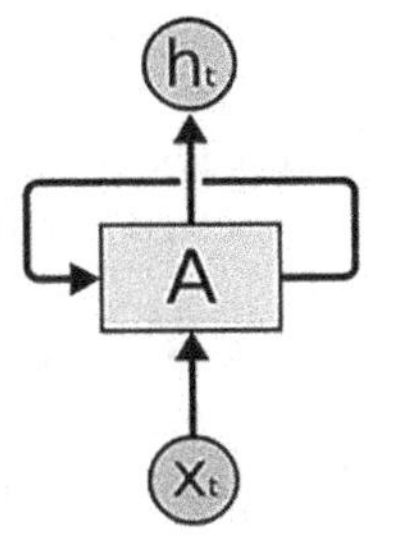

FIGURE 16.1 Recurrent neural network.

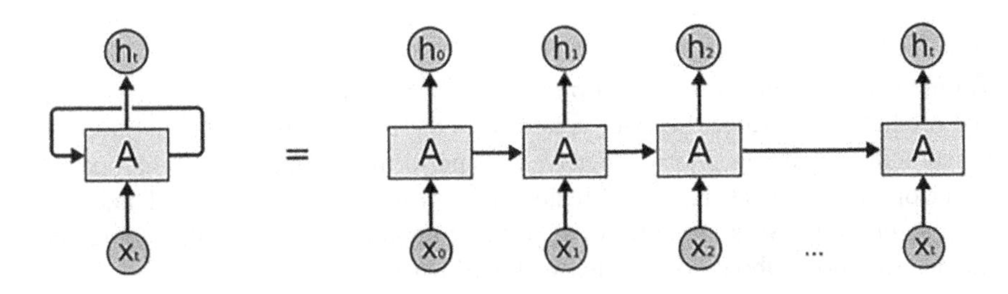

FIGURE 16.2 Unrolled recurrent neural network.

where

h_t = current state
h_{t-1} = previous step
x_t = input state

Formula for activation function:

$$h_t = \tan\text{h}\left(W_{hh}h_t - 1 + W_x h_{xt}\right) \tag{16.3}$$

where

W_{hh} = weight at recurrent neuron
W_{xh} = weight at input neuron

Output of network is calculated with the following formula:

$$y_t = W_{hy}h_t \tag{16.4}$$

where

y_t = output
W_{hy} = weight at output layer

The following steps are taken into consideration while training model through RNN:

1. x_t is provided to the network, which is a single time step of the input.
2. The current state, i.e., h_t, is calculated with the help of current input and previous state.

3. For the next time step, current h_t is considered as h_{t-1}.
4. This procedure is followed as many time steps as the problem demands. Then combine the data from all previous steps.
5. Then, y_t, which is the current final step, is calculated after processing of all the time steps is completed.
6. Finally, the produced output value is compared with the actual output for error calculation.
7. The error value is then backpropagated to the network, and accordingly, weights are updated.

The RNN gets trained in this fashion.

16.2.3.1.1 Problem with Recurrent Neural Network

Training in RNN has three broad parts. Initially, the network does a forward pass and predicts some value. Second, it calculates the loss function. The output of loss function indicated how poorly the network is performing. Finally, it backpropagated the error value to measure the gradients for every node in the network. The internal weights in the network are adjusted depending upon the gradient value. Gradient is the decision point about how the network will learn. If the gradient is big, it means that bigger adjustment is required, and vice versa. This brings the problem that if the gradient to the previous layers is small, then adjustment of the weight to the current layer becomes even smaller. Thus, the gradient shrinks as it passes through the network, and ultimately, the network stops learning. This problem is called as *vanishing gradient* problem. This problem causes RNN not to learn for long-range dependencies over time steps; i.e., RNNs have the short-term memory.

16.2.3.2 LSTM Network Long Short-Term Memory (LSTM)

LSTM neural network is an extension to RNN, which extends its memory. LSTM can learn long-range dependencies. LSTM contains a similar chain-like structure as RNN; the only difference is in the number of neural network layers. LSTM consists of four layers instead of a single layer.

As shown in Figure 16.3, the repeating module in an LSTM contains the four interacting layers. LSTM has gates to protect and control the cell state.

LSTM contains the following components:

1. Forget gate, 2. input gate, and 3. output gate

1. **Forget Gate**: This gate is used to take decision about what should be deleted from the previous state, i.e., $h_{(t-1)}$. The decision is taken after receiving the output of the previous state and the current input xt. LSTM cell visual representation is shown in Figure 16.4.

   It is a sigmoid layer, and it produces the output between 0 and 1.

$$f_t = \sigma\left(w_f\left[h_{t-1}, x_t\right] + b_f\right) \tag{16.5}$$

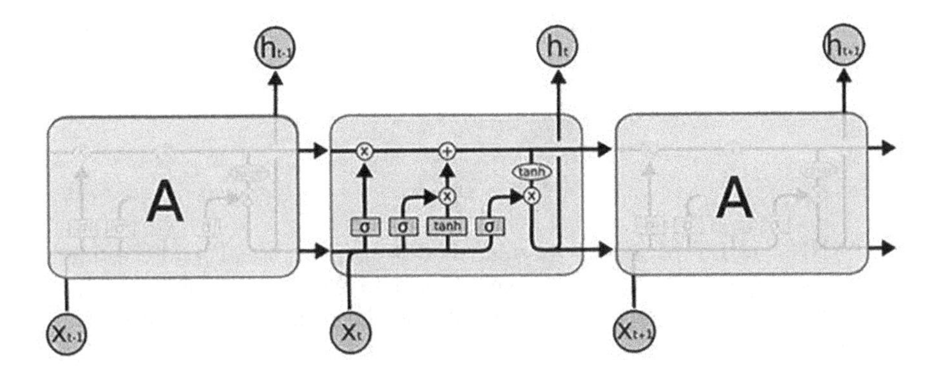

FIGURE 16.3 LSTM with four interacting layers.

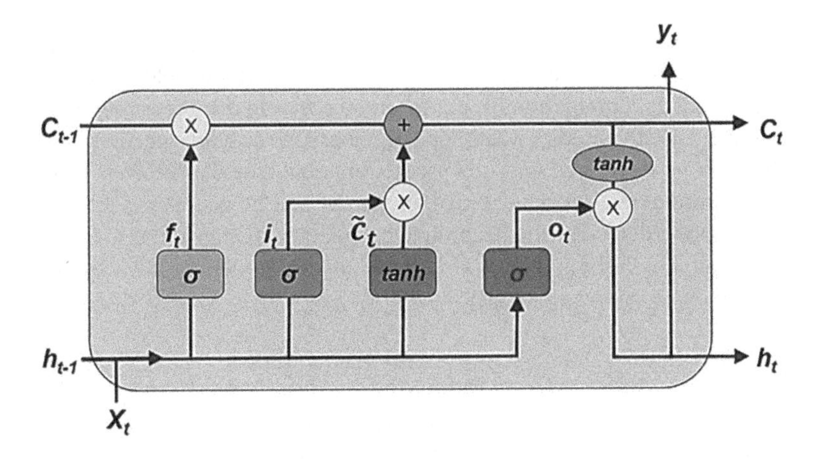

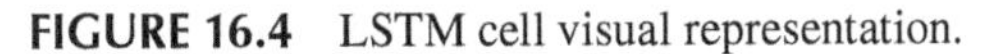

FIGURE 16.4 LSTM cell visual representation.

2. **Input Gate**: Input gate takes the previous output and the next input, and passes them through another sigmoid layer. It returns a value between 0 and 1. The output of this gate is multiplied with the output of the candidate layer.

$$i_t = \sigma\left(w_i\left[h_{t-1}, x_t\right] + b_i\right) \tag{16.6}$$

$$c_t = \tan h\left(W_c * \left[h_{t-1}, x_t\right] + b_c\right) \tag{16.7}$$

This layer applies a hyperbolic tangent to mix the input and the previous output, returning a candidate vector to be added to the internal state.

The internal state is updated with the following formula:

$$c_t = f_t * c_{t-1} + i_t * c_t \tag{16.8}$$

3. **Output Gate**: Output gate controls how much of the internal state is passed to the output.

$$o_t = \sigma\left(w_o\left[h_{t-1}, x_t\right] + b_o\right) \tag{16.9}$$

$$h_t = o_t * \tan h\, c_t \tag{16.10}$$

RNN and LSTM are the popular models used for weather forecasting. However, hybrid models can be adopted to get more accurate results. Additionally, these models are computationally costlier but can be executed in a parallel environment.

16.3 SUMMARY

Getting more timely results depending upon the real-time data is a need of an hour. Nowadays, deep learning techniques like RNN and LSTM are yielding better results for the real-time data prediction. In the future, the hybrid deep learning models can be successfully used to predict weather parameters such as wind speed, humidity etc. Also, they can be integrated with the preexisting applications like Web application, agriculture, production, business decisions, etc. It can be concluded that Artificial intelligence-based deep learning technique can replace human and traditional methods in weather forecasting. The performance of the model can also be increased when implemented on the top of GPUs.

REFERENCES

1. Denis R., Bjarne K.H. 2002. A fuzzy case-based system for weather prediction. *Engineering Intelligent Systems*, 10 (3), pp. 139–146.
2. Guhathakurtha, P. 2006. Long-range monsoon rainfall prediction of 2005 for the districts and sub-division Kerala with artificial neural network. *Current Science*, 90 (6), p. 25.
3. Min J.H., Lee Y.C. 2005. Bankruptcy prediction using support vector machine with optimal choice of kernel function parameters. *Expert Systems with Applications*, 28, pp. 603–614.
4. Mohandes M.A., Halawani T.O., Rehman S., Hussain A.A. 2004. Support vector machines for wind speed prediction. *Renewable Energy*, 29, pp. 939–947.
5. Pal N.R., Pal S., Das J., Majumdar K. 2003. SOFM-MLP: A hybrid neural network for atmospheric temperature prediction. *IEEE Transactions on Geoscience and Remote Sensing*, 41 (12), pp. 2783–2791.
6. Yu P.-S., Chen S.-S., Chang I.F. 2006. Support vector regression for real-time flood stage forecasting. *Journal of Hydrology*, 328, pp. 704–716.
7. Stanislaw O., Konrad G. 2007. Forecasting of daily meteorological pollution using wavelets and support vector machine. *Engineering Applications of Artificial Intelligence*, 20, pp. 745–755.
8. Lu W.-Z., Wang W.-J. 2005. Potential assessment of the support vector machine method in forecasting ambient air pollutant trends. *Chemosphere*, 59, pp. 693–701.

9. Abrahamsen E.B., Brastein O.M., Lie B. 2018. Machine learning in Python for weather forecast based on freely available weather data, *Proceedings of the 59th Conference on Simulation and Modelling (SIMS 59)*, Oslo Metropolitan University, Norway, 26–28 September. doi: 10.3384/ecp18153169.
10. Radhika Y., Shashi M. 2009. Atmospheric temperature prediction using support vector machines. *International Journal of Computer Theory and Engineering*, 1 (1), pp. 1793–8201.
11. Paras M., Mathur S. 2012. A simple weather forecasting model using mathematical regression. *Indian Research Journal of Extension Education Special Issue*, I, pp. 161–168.
12. Sreehari E., Srivastava S. 2018. Prediction of climate variable using multiple linear regression, *2018 4th International Conference on Computing Communication and Automation (ICCCA)*, Greater Noida, India, pp. 1–4, doi: 10.1109/CCAA.2018.8777452.
13. Amral N., Ozveren C.S., King D. 2007. Short term load forecasting using multiple linear regression, *42nd International Universities Power Engineering Conference*, Brighton, pp. 1192–1198, doi: 10.1109/UPEC.2007.4469121.
14. Kumar A., Singha M.P., Saswata G., Abhishek A. 2012. Weather forecasting model using artificial neural network. *Procedia Technology*, 4, pp. 311–318.
15. Chattopadhyay S., Chattopadhyay G. 2008. Identification of the best hidden layer size for three-layered neural net in predicting monsoon rainfall in India. *Journal of Hydroinformatics*, 10 (2), pp. 181–188.
16. Tamboli S., Bewoor L. 2018. A review of soft computing technique for real-time data forecasting, *International Conference on Communication and Information Processing (ICCIP-2019)*, pp. 1–7, doi: 10.2139/ssrn.3418746.

17 Molecular Mining

Applications in Pharmaceutical Sciences

Anurag Verma, Ashutosh Rathore, Shalini Tripathi and Pardeep Kumar Sharma
Lovely Professional University

CONTENTS

17.1 INTRODUCTION

The increasing amount and complexity of data calls for new and flexible approaches to mine the data. The traditional method of manual data analysis has now become inefficient, and computer-based analysis has now become necessary. Statistical methods, expert systems, fuzzy neural networks, and machine learning (ML) algorithms are extensively studied and applied to data mining.

Drug plays a very much essential role in everyone's life. To produce any new drug for the treatment of a disease and to launch it for public interest using conventional methods is very long running/tiresome and very uneconomical. Several molecules are combined to form a new drug, so it is necessary to find the structure of molecules, which can be done through data mining. Data mining means finding a new information from a lot of biological/structural data that has been previously given to the system, and information which is obtained is new and useful.

Drug is a chemical stuff that is mainly used for recognition/identification or therapy of an illness/disease. Diseases are recognized at the subatomic level called as target particle. Any illness is treated by counterbalancing the target particle interacting with the active drug molecule. For different diseases, there are various targets as well as many active drug compounds. So, it becomes very necessary to classify or identify active chemical compounds from a group of large biochemical information/statistics, which comprises active, inactive, and inconclusive compounds. Practically, designing of a new drug involves processes such as screening of chemical compounds as well as categorizing those compounds, which is very noneconomical in terms of cost (average cost requires approximately US $ 1 billion) and time (around 10–15 years). Therefore, drug design is the foremost and demanding piece of work in clinical research.

Data mining means to examine the preexisting large databases to discover/extract something which is new and useful to us. Like data mining, molecular mining involves discovering new molecules from large biological data sets. The new molecules that are discovered can hit the target more efficiently and hence have a better therapeutic efficacy.

Most of the data mining tasks in bioinformatics comprise scanning massive groups of molecules with an objective to discover some uniformity among molecules of a specific class. An example for the same is drug discovery, where the scientist wishes to discover a new drug candidate based on the exploratory confirmation of activity against a certain illness gathered by scanning several thousands of molecules. The present-day emphasis comes from chemical synthesis success prognosis, where the major aim is to find the molecular attributes that hinder/obstruct the desired reaction.

17.2 WHY MOLECULAR MINING?

To produce any new drug for the treatment of a particular disease and launch it for public interest using conventional methods is very long running/tiresome and uneconomical. Several molecules are combined together to form a new drug, so it is necessary to find the structure of molecules, which can be done through data mining. Data mining means finding a new information from a lot of biological/structural data that has been previously given to the system, and information which is obtained is new and useful.

17.3 TOOLS INVOLVED IN DATA MINING

Data mining involves the basic three tools:

1. Statistics
2. Artificial intelligence (AI)
3. Machine learning

Statistics means a large preexisting biological data set having n number of molecular structures from which a new and useful data is discovered. In other words, we can say that it is a type of mathematical analysis, which involves the use of quantified models for any set of data. Statistics uses methodologies to extract, analyze, revise, and make a conclusion from the given data.

In computer science, AI is commonly known as machine insight, i.e., knowledge exhibited by machines, rather than the normal insight shown by people. It may also be defined as a capacity of an advanced PC or PC-controlled robot to perform assignments generally connected with intelligent beings.

Machine learning is the technical learning of algorithms and statistical representation that PC utilizes to carry out a particular piece of work without making use of direct/clear-cut commands, depending on models and inferences instead. It is viewed as a subdivision of AI.

17.4 DATA SCIENCE

Before going in detail to AI, let us first learn about data science. Data science is a broad term that covers all aspects of data processing and not only an analytical but also an algorithmic angle. It includes the following aspects:

1. Data visualization
   This is the effort/means of placing the data in a visualized format for a better understanding.
2. Data integration
   This means collecting data from different sources and then combining it in a single format. Integration is done with the help of the following processes like cleansing, mapping, and transformation.
3. Distributed architecture
   This includes models, policies, rules, and standards, which help in governing what type of data is grouped, in which manner it is piled, presented, combined, and set to employ in information systems and in organizations.
4. Data-driven decisions
   This is a method to business governance that assess choices that can be retreated with provable stats.
5. Automation using ML
   This is the primary way to automate the data by the approach of ML.
6. Data engineering
   This aspect of data sciences mainly focuses on practical applications of data collection and analysis.

17.5 MACHINE LEARNING

Machine learning is a piece of AI that educates/trains a PC program or calculates the capacity to naturally gain proficiency with an errand and develop itself consistently for a fact. For the most part, it centers around the improvement of PC programs which can learn and get to information for themselves. Developers need to code and inspect cautiously as indicated by the need with the goal that framework can autonomously perform the visit enhancements. The process of ML is shown in Figure 17.1.

There are three kinds of ML:

- Unsupervised learning (Figure 17.2)
- Supervised learning (Figure 17.3)
- Reinforcement learning (Figure 17.4)

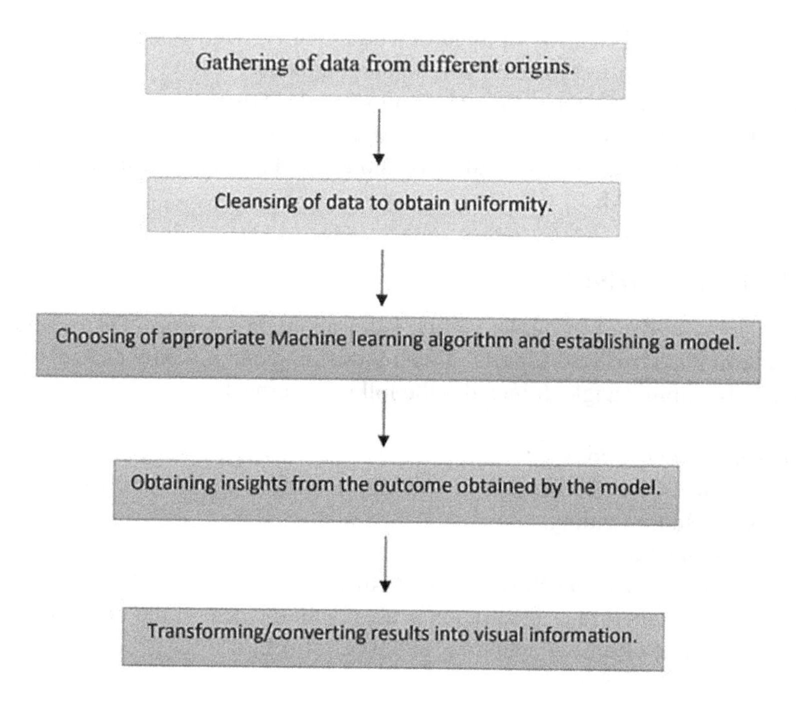

FIGURE 17.1 The process of machine learning.

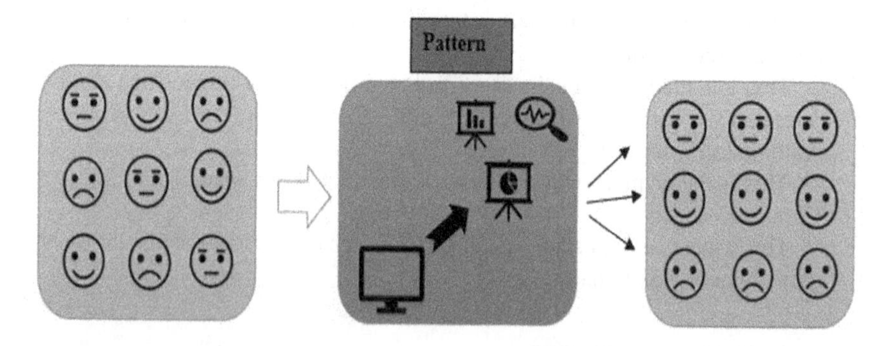

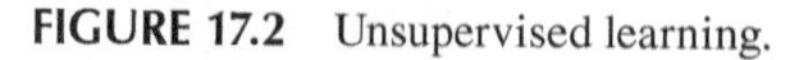

FIGURE 17.2 Unsupervised learning.

17.6 ML TECHNIQUES

Various ML tools are available for the classification of different sets of data. The ML tools that find a role in pharmaceutical drug design are neural network, k-NN (k-nearest neighbor), support vector machine (SVM), Naive Bayes (NB), and decision tree (DT). The best ML tool that can be used further in the drug design can be found by applying different matrix systems to these ML tools, and the results can be interpreted and compared.

Artificial neural network (ANN), NB, SVM, k-NN, DT, logistic regression (LR), linear discriminant analysis (LDA), and Random Forest (RF) have different

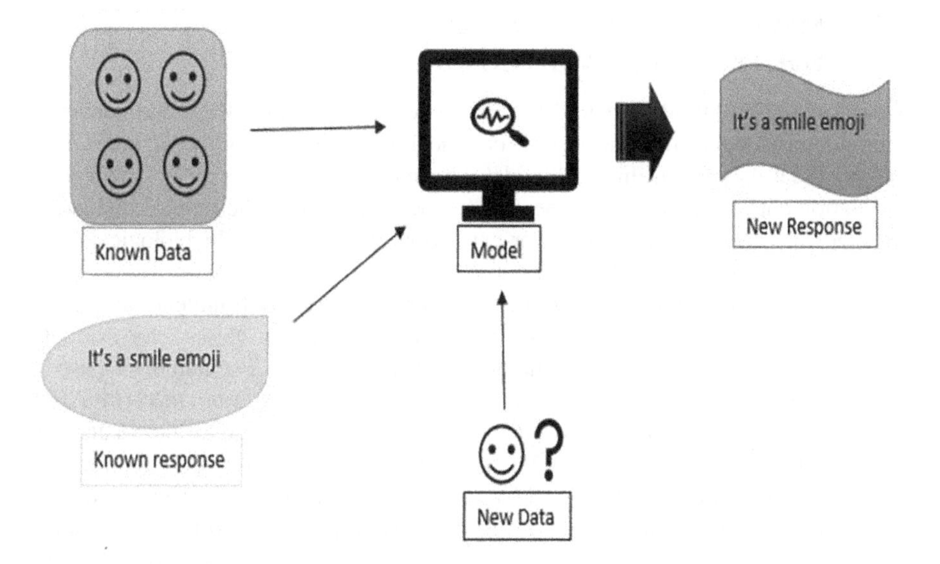

FIGURE 17.3 Supervised learning.

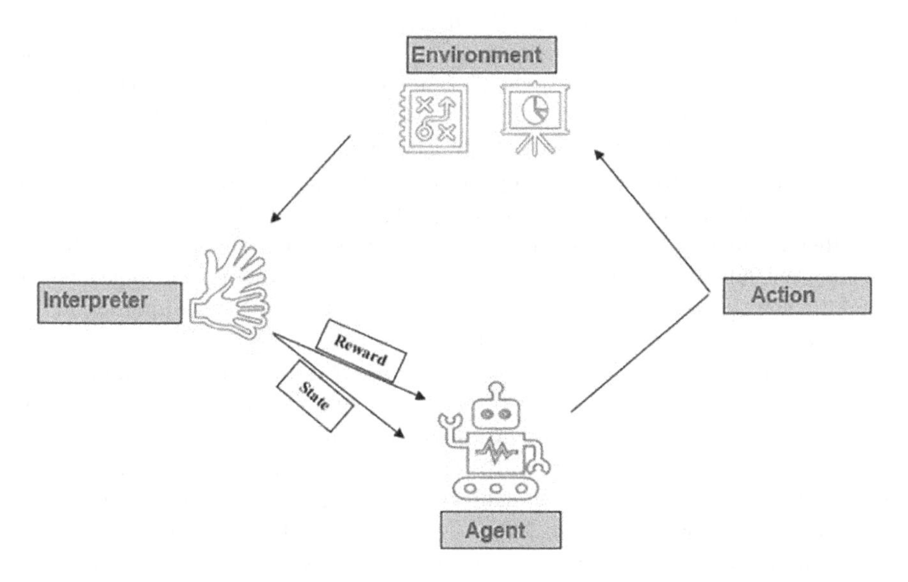

FIGURE 17.4 Reinforcement learning.

algorithms on which they work. Any single ML tool or simultaneous two tools can be trained by giving a data set. Different matrix systems such as confusion, accuracy, recall, *F*-1 score, and precision are applied to our trained program to get the desired results. By comparing the results of two simultaneous algorithms, one can easily identify which ML tool is a better classifier and which classifier suits best to this study.

17.7 MACHINE LEARNING APPROACHES FOR MINING OF MOLECULES

Geerestein (2000) used a DT to differentiate between potential drugs and nondrugs. Compounds taken from the accessible chemical index and the world drug index databases were utilized for training of model. The inaccuracy measure on an independent validation data set was found to be 17.4%. The prediction of the model can also be utilized to escort the procurement or extraction of compounds for biological scanning or making of combinatorial libraries.

Christian Borgelt (2002) utilized an algorithm to discover fragments in a group of molecules that assist in distinguishing between dissimilar classes, for example, activity that would further aid in the field of drug discovery. The author has used a data set, which is available publicly from National Cancer Institute, the DTP AIDS Antiviral Screen data set. In this scanning, the author utilized the formazan analysis to estimate the conservation of human CEM cells from HIV-1. Compounds having 50% conservation/protection to CEM cells were experimented again and were recorded as moderately active (CM). And compounds providing 100% conservation/protection were recorded as confirmed active (CA). And the compounds that do not fulfill any of these criteria were recorded as confirmed inactive (CI). A total of 37,171 compounds were tested, and it was found that 325 compounds belong to CA, 877 compounds belong to CM, and 35,969 compounds belong to CI.

(Serra & Thompson (2003) generated the classification models for the prediction of in vitro cytogenetic result for a group of 383 organic compounds. Two techniques k-NN and SVM were utilized. Values were taken from the assay that was carried on hamster lung cells, which includes both 24- and 48-hour exposure. Different descriptors were used to encode the topological, electronic, geometrical, or polar surface area features of the structure. The final categorization success percentage for a k-NN classifier assembled with only 6 topological descriptors was found to be 81.2% for the training set and 86.5% for the testing set. The total categorization success percentage for a three-descriptor SVM model was found to be 99.7% for the training set, 92.1% for the cross-validation set, and 83.8% for the testing set.

Evgeny Byvatov (2003) compared SVM and ANN systems for drug/nondrug classification. The author applied both the systems for classifying compounds data set into drug and nondrug or to filter potentially unwanted molecules from a compound library. A total of 9208 molecules were taken into consideration (4998 drugs and 4210 nondrug molecules). Execution rates of both classifiers were compared using different descriptors—120 standard Ghose-Crippen fragment descriptors, and a broad variety of 180 different properties and physicochemical descriptors from the Molecular Operating Environment (MOE) package, and 225 topological pharmacophore (CATS) descriptors. A total of 525 descriptors were taken into consideration to cross-validate the results, and it was found that the results obtained by SVM were more accurate (82% correct predictions) and that of ANN was found to be 80%.

Andreas Bender (2004) introduced a new way to search similarity. In this technique, molecules were being entitled as an atom environment, which were then fed into an information gain-based system. Then, a NB classifier is employed for the classification of compounds. The algorithm surpasses all current recovery techniques assessed here using two- and three-dimensional descriptors. This technique can also be utilized for the recognition of functional groups in active molecules and is computationally effective.

Gongde and Neagu (2005) proposed a vigorous technique/procedure, fuzzy k-NN model, for the estimation of toxicity in chemical compounds. The method was basically dependent on the supervised clustering method, known as k-NN model, which works on fuzzy segregation/clustering instead of crisp segregation/clustering. The exploratory results of fuzzy k-NN model supervised on 13 public data sets from UCI machine learning repository and seven toxicity data sets from the real-world applications were contrasted with the outcomes of fuzzy c-means clustering, k-means clustering, k-NN, fuzzy k-NN in terms of execution based upon the classification. The results showed that the fuzzy k-NN model was the best technique for the toxicity prediction of chemical compounds.

Shubhangi & Hiremath (2008) used SVM with neural network for a handwritten character recognition. The character recognition technique is either a printed image/document or a handwritten character recognition. They showed a cooperation of combining SVM classifiers with neural network using the morphological feature set. An approach was proposed that can be used to detect and recognize both the categories of the characters (digital and hand-written). And it was found that the classifiers have classified correctly up to 97%.

Mandal & Jana (2013) performed a relative investigation of NB and k-NN algorithm for the multiclass drug molecule categorization. NB and k-NN algorithm exploration was performed on the biochemical record of data, which were taken from PubChem. A data set (i.e., qHTS assessment to check autofluorescence of compound at 460 nm (grey) in HEK293 cells) was taken into consideration for the analysis, which comprises a total of 1280 records where each record is affiliated to active, inactive, or inconclusive group. From the data set, 20% of data was kept for the examining purpose and the rest (80%) was utilized for the training of model. Testing data set comprised a total of 256 compounds, and out of them, 14 were active, 54 were inconclusive, and the rest of the compounds were inactive. The NB classifier miscategorized only four compounds (one compound was anticipated as active but actually that was inactive, and three compounds were anticipated as active but actually those were inconclusive). The k-NN classifier miscategorized only one compound (anticipated as inconclusive but actually that was active). The accuracy of NB classifier was found to be 93%, and that of k-NN classifier was found to be 99.6%.

Ioannis et al. (2017) used ML and data mining methods in an ongoing research of diabetes for transforming all intelligently available data into valuable knowledge. Substantial exploration in all aspects linked/affiliated to diabetes (such as diagnosis, etio-pathophysiology, treatment etc.) has escorted to the collection of massive amounts of data. The main objective of this research was to scrutinize all ML and data mining techniques in the field of diabetes experimentation with respect to all of the above aspects. They employed a broad variety of ML algorithms. Mainly, among all, approx. 85% were characterized by supervised learning approaches, and the rest (15%) with the help of unsupervised learning, i.e., by association rules. And they found that SVM was the most successful and widely used algorithm.

Lei Zhang (2018) developed a PC-helped atomic plan/screening strategy for structure and screening of scent particles. The author used a method by which the essence of particles was anticipated using a ML technique, and a group contribution mechanism was employed for anticipating physical attributes like solubility parameter, vapor pressure, and viscosity. A MILP/MINLP model was made for screening/depiction of scent particles. And the results showed that even the minor changes in the molecular structure may result in exceptionally different aromas. Thus, molecular portrayal having larger sets of structural framework needs to be developed to establish a model that can anticipate more precisely. And it was also found that molecules with homogeneous vibration spectrum have homogeneous aroma properties.

Konstantinos Vougas (2019) has described a new in silico scanning procedure basically dependent on the association rule mining, to recognize genes as individual operators of drug response and to contrast them with appropriate data mining techniques. The author developed a computer-based model for anticipating biological results involving mainly three steps: picking of data set, selection of the right algorithm and instructing it to develop an anticipation model, and testing it in new data sets.

17.8 PROCEDURE

A model was prepared for comparing different ML tools for the classification of compounds.

For practical analysis, laptop descriptions are Intel core i5 7th gen with 8 GB Random Access Memory (RAM) having Ubuntu (18.04.3 LTS) operating System. "Python" (version 3.7.3) was used as a programming language for training and testing of different models.

The main theme of this model was to classify/categorize chemical compounds into active, inactive, and inconclusive. Comparison was done between k-NN, SVM, DT, LR, LDA, RF, and NB. All standard algorithms were used except in NB (Gaussian NB was used). Metrics systems that are used here for performance analysis are precision, recall, and F1 score.

A chemical compound data set was taken from PubChem (AID 720678). This bioassay record (AID 720678) is associated with a total of 98 additional bioassay records in PubChem, which includes several assay projects (https://pubchem.ncbi.nlm.nih.gov/bioassay/720678).

A total of 10,486 compounds were there in the data set. Of them, 10,374 were inactive, 40 were active, and 72 were inconclusive. Split format (70/30) was used; i.e., 70% of the total data set was used for the training of model and the remaining 30% for the testing purpose. ML library sklearn was used in Python. Classification results of different models are summarized in Table 17.1. The score data of each model is summarized in Table 17.2.

TABLE 17.1
Classification of Different Models

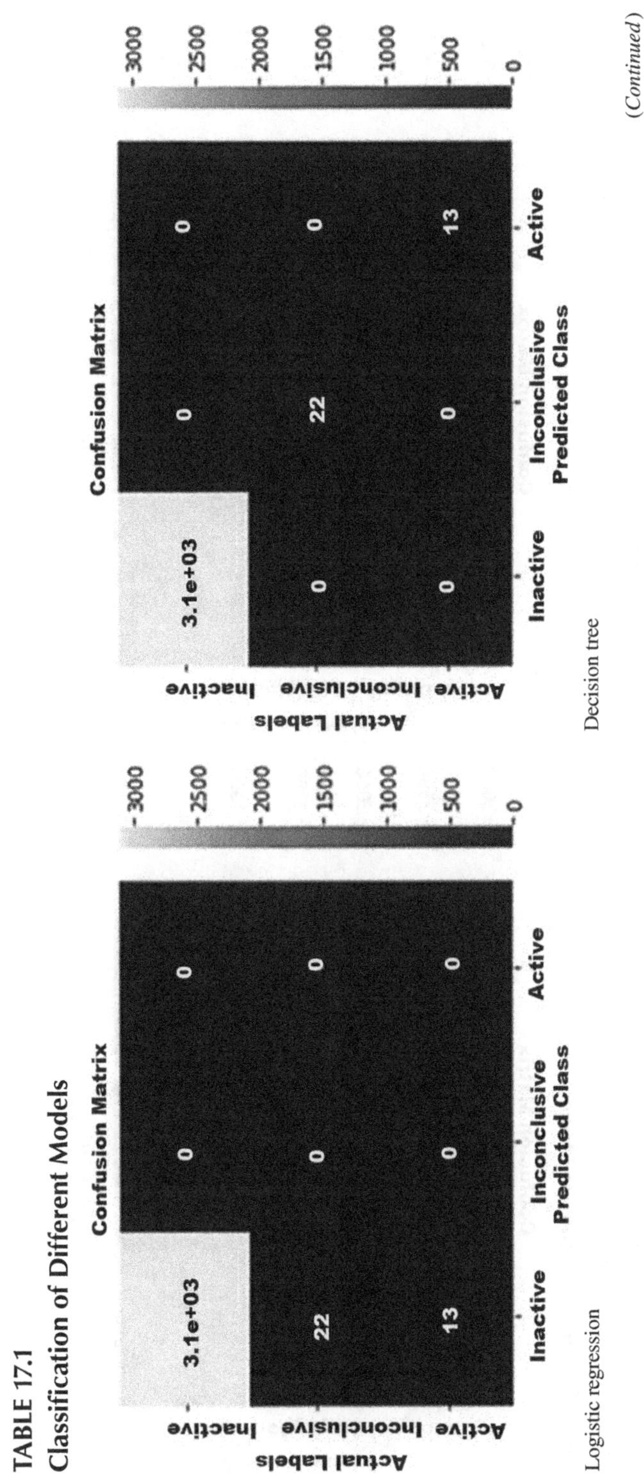

Logistic regression

Decision tree

(Continued)

TABLE 17.1 *(Continued)*
Classification of Different Models

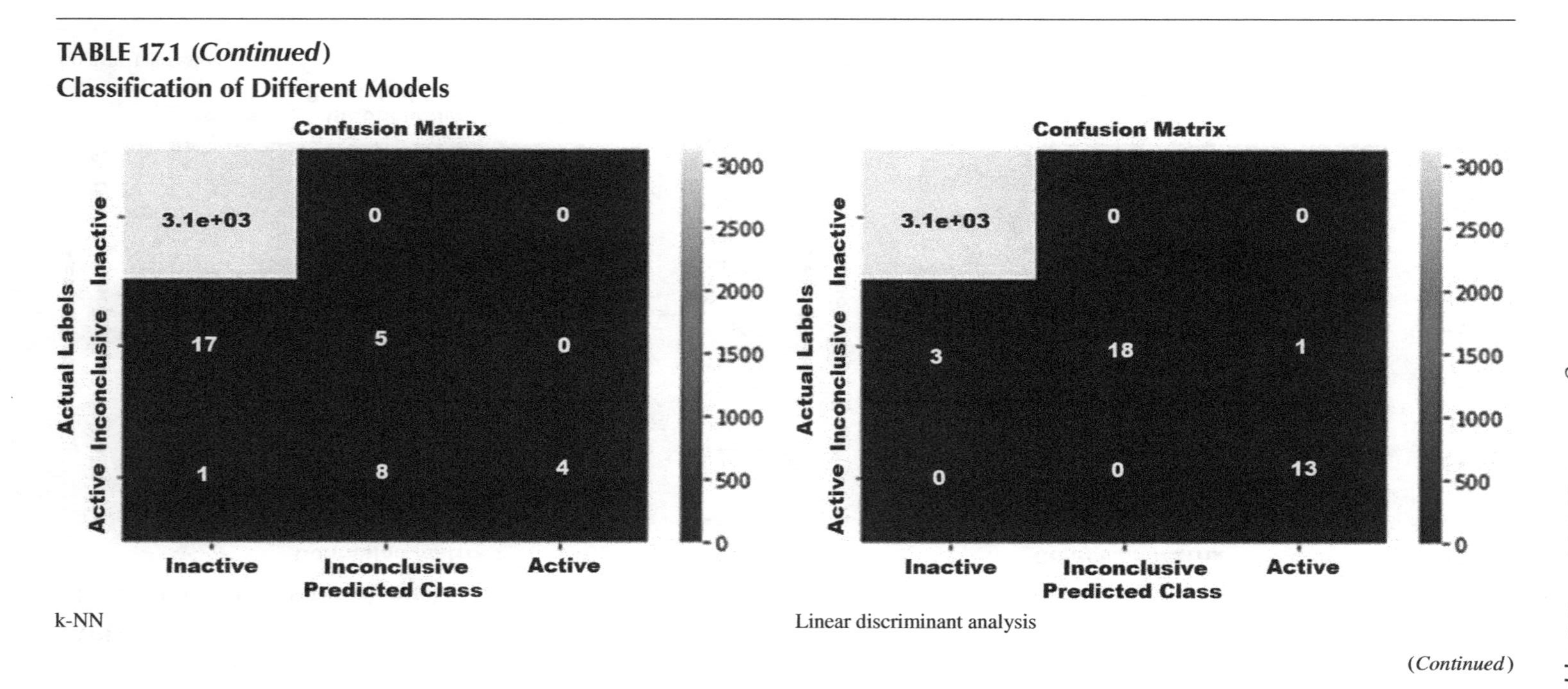

k-NN

Linear discriminant analysis

(Continued)

TABLE 17.1 (*Continued*)
Classification of Different Models

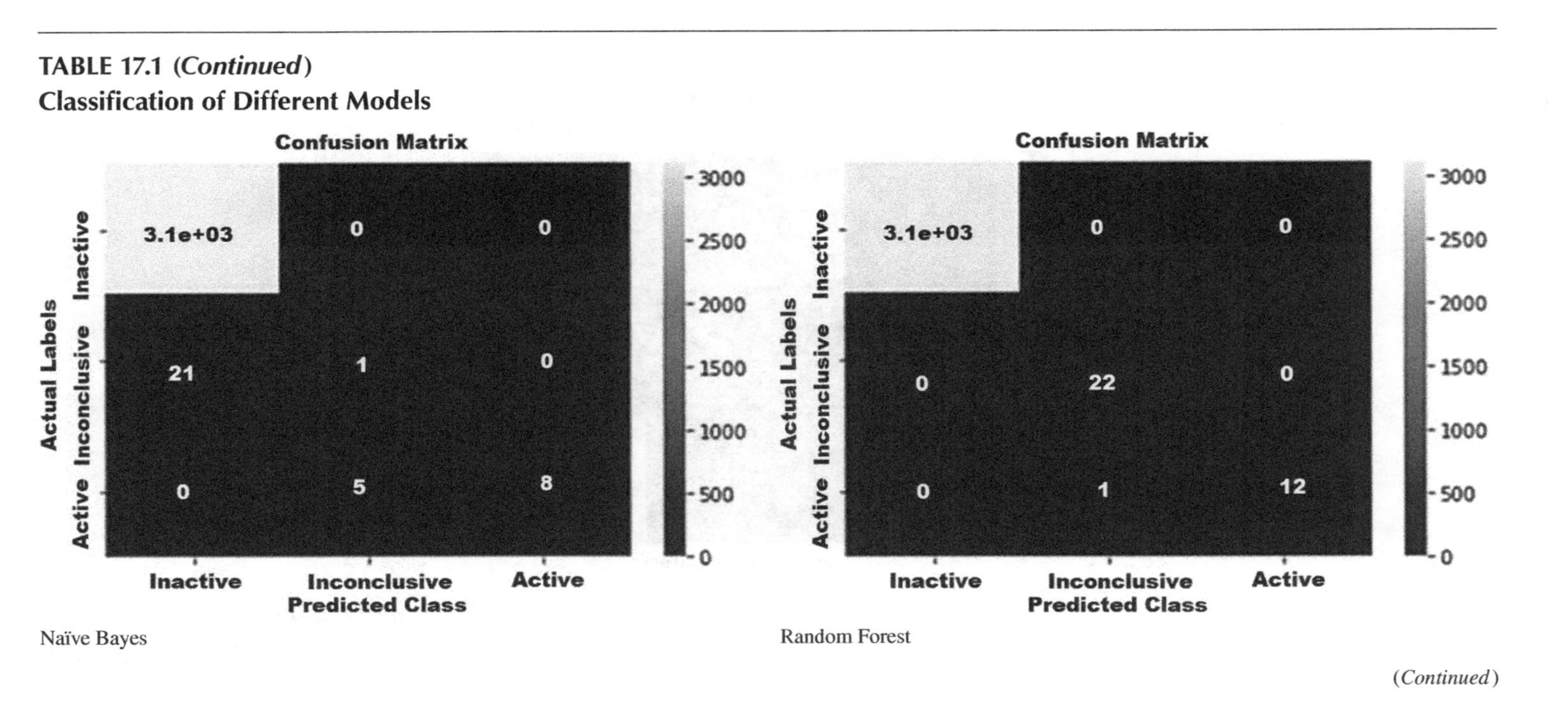

Naïve Bayes

Random Forest

(*Continued*)

TABLE 17.1 (*Continued*)
Classification of Different Models

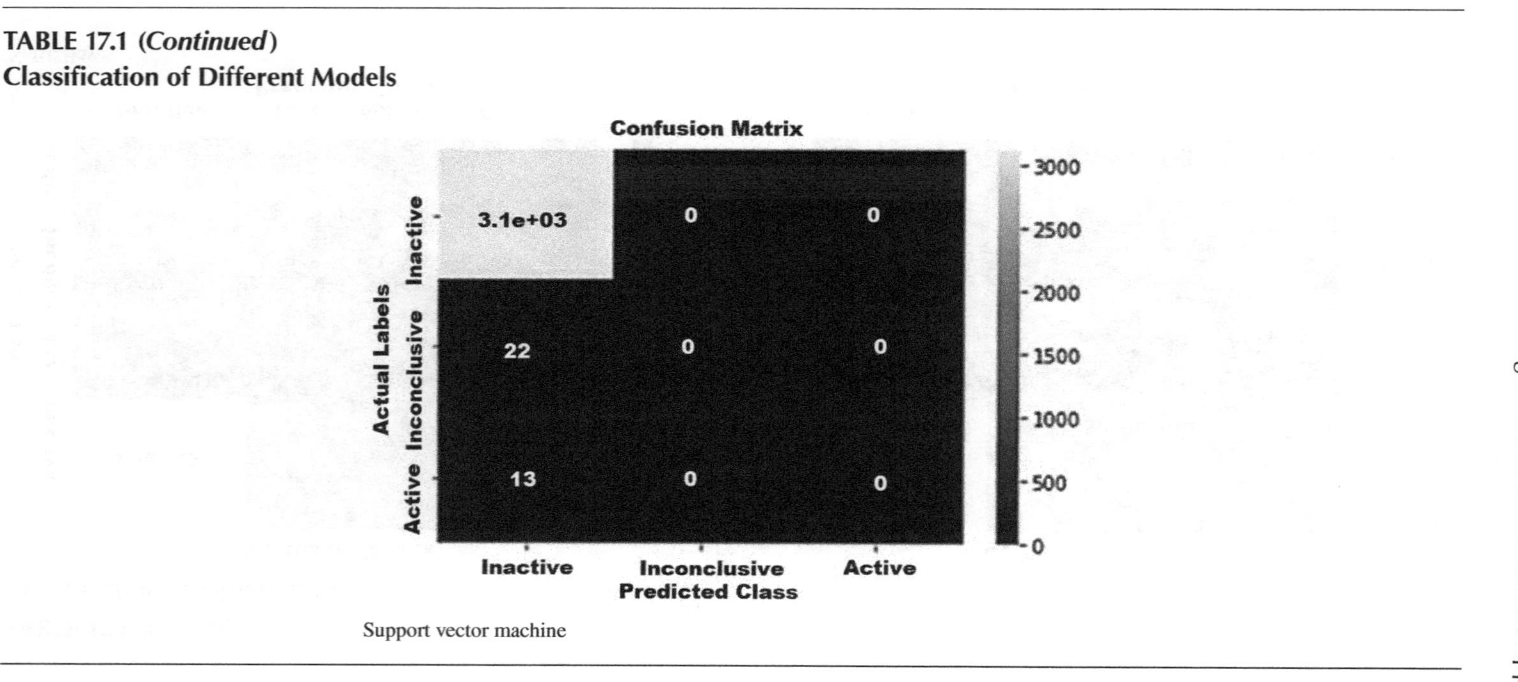

Support vector machine

TABLE 17.2
The Score Data of Each Model

ML Techniques	Avg Precision Score (%)	Avg Recall Score (%)	Avg F1 Score (%)
Logistic regression	32.96	33.33	33.14
Decision tree	100	100	100
k-NN	79.29	51.16	58.44
Linear discriminant analysis	97.58	93.93	95.41
Naïve Bayes	89.88	85.66	86.74
Random Forest	98.55	97.43	97.92
Support vector machine	32.96	33.33	33.14

17.9 CONCLUSION

Training and testing of different ML techniques were done using the data set mentioned above, and their results were compared using three metrics systems, and it was found that DT is the most effective technique with 100% accurate results and SVM and LR are the least effective techniques with just 33% accurate results. And other than these techniques, LDA and RF also provide the promising results.

REFERENCES

Andreas Bender, H. Y. (2004). Molecular Similarity Searching Using Atom Environments, Information-Based Feature Selection, and a Naïve Bayesian Classifier. *Journal of Chemical Information and Computer Sciences*, *44*(1), 170–178.

Christian Borgelt, M. R. (2002). Mining Molecular Fragments: Finding Relevant Substructures of Molecules. *IEEE Transactions on Neural Networks and Learning Systems*, 51–58.

Evgeny Byvatov, U. F. (2003). Comparison of Support Vector Machine and Artificial Neural Network Systems for Drug/Non-Drug Classification. *Journal of Cheminformatics*, 1–7.

Geerestein, M. W. (2000). Potential Drugs and Nondrugs: Prediction and Identification of Important Structural Features. *Journal of Chemical Information and Computer Scientists*, 280–292.

Gongde, G., & Neagu, D. (2005). Fuzzy kNN Model Applied to Predictive Toxicology Data Mining. *International Journal of Computational Intelligence and Applications*, *5*(3), 321–333.

Ioannis, K., Olga, T., Athanasios, S., Nicos, M., Ioannis, V., & Chouvarda, I. (2017). Machine Learning and Data Mining Methods in Diabetes Research. *Computational and Structural Biotechnology Journal*, *15*, 104–116.

Konstantinos Vougas, T. S.-R. (2019). Machine Learning and Data Mining Frameworks for Predicting Drug Response in Cancer. *Pharmacology & Therapeutics*, 2–28.

Lei Zhang, H. M. (2018). A Machine Learning Based Computer-Aided Molecular Design/Screening Methodology for Fragrance Molecules. *Computers and Chemical Engineering*, 2–47.

Mandal, L., & Jana, N. D. (2013). A Comparative Study of Naïve Bayes and K-NN Algorithm for Multi-Class Drug Molecule Classification. *International Journal of Computer science and Applications*, 1–6.

Serra, J. R., & Thompson, E. D. (2003). Development of Binary Classification of Structural Chromosome Aberrations for a Diverse Set of Organic Compounds from Molecular Structure. *Chemical Research in Toxicology*, *16*, 153–163.

Shubhangi, D. C., & Hiremath, P. H. (2008). Multi-Class SVM Classifier with Neural Network for Handwritten Character Recognition. *International Journal of Computer Science and Applications*, 2–5.

Index